This book is *Introducing How the Solar System Forms a Pre-Script*

by for the first time ever being able to

Explain existing issues about the Introducing of the working application of

The Titius Bode Law

ISBN-13: 978-1507774205
(CreateSpace-Assigned)
ISBN-10: 1507774206

WRITTEN BY Peet (P S. J.) SCHUTTE

©KOSMOLOGIESE EN ASTRONOMIESE TEGNIKA

http://www.titius-bode-law-explain.co.za/index.html

This is to introduce the following books:

An Academic Introducing to The Titius Bode Law Book 1

An Academic Introducing to The Titius Bode Law Book 2

An Academic Introducing to The Titius Bode Law Book 3

And also

A Cosmlc Birth as an Academic Presentation Book 1

A Cosmic Birth...as a Special Presentation Book 2

The above mentioned books are only available from the CraeteSpace website

This book will introduce you to the
Absolute Relativity
Of Singularity

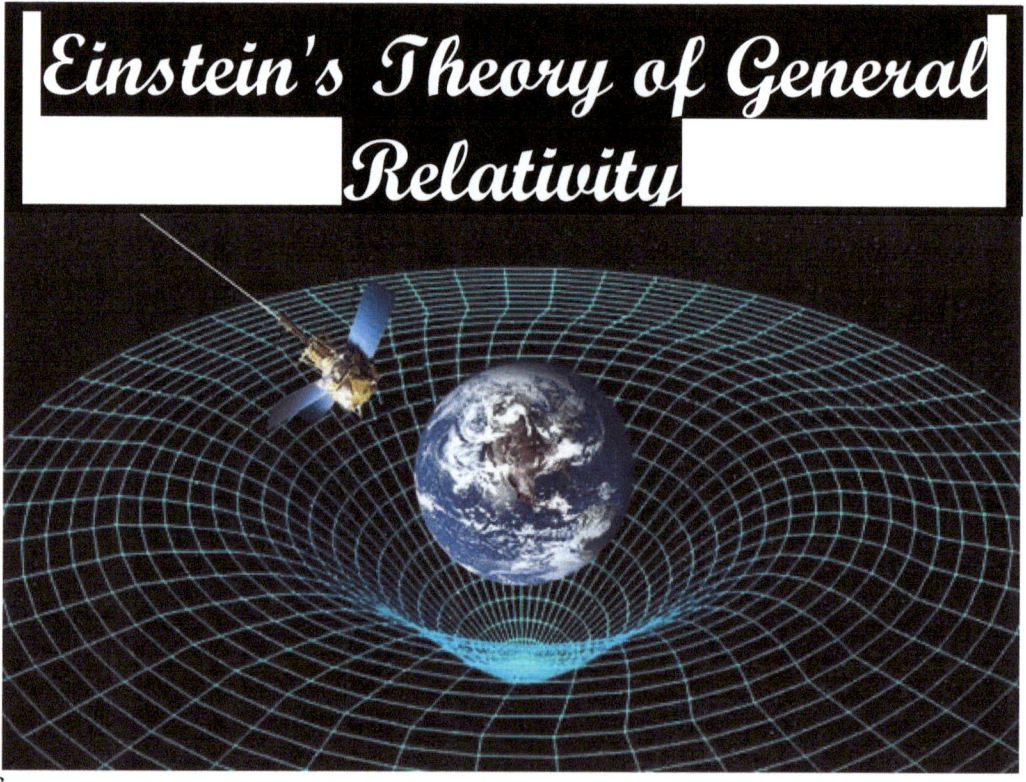

Einstein's Theory of General Relativity

Taken form Space .com

In 1905, Albert Einstein determined that the laws of physics are the same for all non-accelerating observers, and that the speed of light in a vacuum was independent of the motion of all observers. This was the theory of special relativity. It introduced a new framework for all of physics and proposed new concepts of space and time.

Einstein then spent 10 years trying to include acceleration in the theory and published his theory of general relativity in 1915. In it, he determined that massive objects cause a distortion in space-time, which is felt as gravity.

Now I am introducing you to the concept, which I prove as the Concept I named the Absolute Relativity of Singularity.

There are 4 laws in the cosmos that I was able to decipher and from investigating these laws it enabled me to find the centre of the Universe

Whether you accept this statement or not, but for the past three hundred years there is a Conspiracy in Science in Progress and moreover it has been going on as a deliberate result of science brainwashing students in accepting absolute bogus information as truthful. Those in science pretend Isaac Newton's principles work and physicists make Newtonian principles work notwithstanding nature showing the very opposite. Nothing about what Newton says vaguely corresponds with what applies in nature! Scientists forcefully present science as a copy of nature but that misrepresents the truth. If you read only this website you will come to see what Newtonian science could never explain. It's called the Titius Bode law. For the first time ever I explain this law.

The solar system does not use mass to form gravity as Newtonian science declares but this information is never often and openly revealed to the public as detrimentally important. Nature does not apply Newton and Newtonian science but uses another application going by the name of the **Titius Bode law**.

Go on and look it up on the Internet and verify what I say about nature using the **Titius Bode law** being applied instead of Newton. This is fact not widely promoted but nevertheless it is true.

In four books on different levels of intensity in each I show and explain just how I cracked the principle named the **Titius Bode law** and show why it applies and how nature works and in each one of four books on a different understanding level I reveal the concept that **nature** (**not Newton or science**) uses as the building blocks of space. But the concept disproves Newtonian science and rubbishes Newton's ideas about what he thought gravity is as unworkable and it is that part that science will never accept as a factual scientific fact. **Nature dismisses Newton's mass ideas**. I have been fighting this fraud for decades but I found that there is too much money to be lost to recognise my work and too much honour going wasted if mainstream science had to freely admit that in their conduct they are fraudsters, which is exactly what they are. Download any of the four books I give for free and test what I have to say. The books I offer as an introduction is free but it delves into science in a manner you have never experienced before. However, I show what is incorrect with science but if you wish to read the corrections you have to purchase. Download the free books and see how revealing the information is on how science cheats.

The fact that nature uses the Titius bode law in the solar layout is known for centuries but as usual science tells the Universe what it is and ignores what the Universe really is. Science ignores nature since science has no idea why the Universe is what it is. The solar system and therefore the entire Universe uses a ratio that two men Titius and Bode simultaneously but being well apart at the time discovered. The solar system and indeed the Universe is built not by "mass" as Newton suggested incorrectly but employing another unknown system but is also a very well hidden system called the Titius Bode law. This law is what forms the solar system proving it is not mass. **The following table compares the law's predictions with the actual distances, where the addition of Pluto is a modern modification.**

Planet	n	Titius-Bode Law	Semi-Major Axis
Mercury		0.40	0.39
Venus	0	0.70	0.72
Earth	1	1.00	1.00
Mars	2	1.60	1.52
asterold belt	3	2.80	2.8
Jupiter	4	5.20	5.20
Saturn	5	10.0	9.54
Uranus	6	19.6	19.2
Neptune	-	-	30.1
Pluto	7	38.8	39.4

This is what is there: It is the **Titius Bode Law** and I show why the **Bode's Law"** or **"Titius-Bode Law"** forms this formation ratio as it does but the best science could come up with was to change the original formulation that was $a = (n + 4) / 10$ where $n = 0, 3, 6, 12, 24, 48...$ to the modern formulation of ($AU_{earth} = 147.597 *10^6$ km): $a = 0.4 + 0.3 \times k$ where "k'= 0, 1, 2, 4, 8, 16, 32, 64, 128 (sequence)

The ratio is so impressively periodic or cyclic correct that it can be put to a formula such as $a = (n + 4)/10$ and the outcome is so predictable that according to the formula $a = (n + 4)/10$ it led to all the discovery of all the missing planets that was discovered after Galileo Galilee used the first spyglass to look at stars. This ratio doubles the distance a planet has every time a new planet is located in a new position. The distance of Saturn doubles the distance between it and Jupiter to what the distance is between Jupiter and the sun. The distance ratio doubles to Mars to what the distance is from the sun to the earth. This ratio of doubling the relevancy in distance from the sun applies notwithstanding the size or the mass or any specific value that any planet might have. This shows that the ratio is not vested in material but it is in accordance with the space that holds the material and that throws "mass" and Newton out of the window forever. This law has been and is known to science for almost if not two hundred years and if you are not a professional astrophysicist you have never heard of this law before in your entire life. It is well hidden under Newton's misconceptions and no matter how much any one wishes to share some brainwashing therapy with me the fact is that Newton's $F = G \dfrac{M_1 M_2}{r^2}$ and $4\pi^2 a^3 = P^2 G(M + m)$ does not wash because it is not applying in any form within the Universe anywhere. The Titius Bode law is what the solar system uses and if that burns your brainwashing and you mind control inflicts excruciating pain on your brain and you get an overwhelming urge to becomes violent with disapproval towards me it would be best to kneel and pray for it is not me that put the Titius Bode law in application within the cosmos but it is the way God Almighty designed the Universe. When you argue with me you have an issue with God or nature or the cosmos, not with me. It is used in nature and not by me 'cause I show it's there and that's all. What I achieved is to find out why the Titius Bode law applies and how does it come about to form the ratio that it does. I was the one that made the connection (not discovery) how gravity comes about by the implication of the Titius Bode law but also always in conjunction with three other phenomena called
1) The Coanda effect, which is the way, the atmosphere forms.
2) The Titius Bode law is how planets use a ratio to arrange their allocated positions.
3) The Roche limit is the law that applies to what we call as the "sound barrier" and how stars explode.
4) The Lagrangian points is why atmospheric layers form around the earth and some planets has rings.

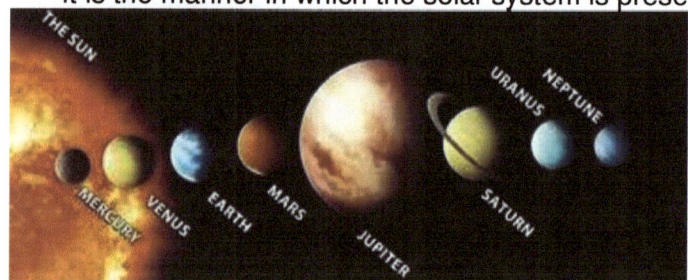

It is the manner in which the solar system is presented that is completely inaccurate and just as much confusing. The solar system presented with this layout is even less accurate than what the Ptolemaic presentation is because the Ptolemaic have the numeric order correct and the rest is completely mythical. Having the solar system is so close in proximity gives the impression of being cramped for space and huge in material and there is no excuse to explain away the inaccuracy except to help Newton cheat by cheating some more. The distances apart are more important to science than what sizes are. Newtonian science dismisses that the sun spins around the earth but that I observe every day of my life. In this time I can see the stars in constellations determining which month it is and what season we are in. The stars turn around the earth just as the moon and the sun and denying this is misrepresenting truth.

By investigating the four laws that form the Universe I was able to locate the centre of the Universe. This is no hoax because I am able to point your finger on the centre spot of the Universe. I located the very first point from where the Universe started and that is how I could follow the process how the Universe started and from there developed. Every scientist will tell you that finding the point where the Universe started and the point forming the centre of the Universe is impossible. They also form the opinion that the Titius Bode law can't mathematically be explained. It can't if you use Newtonian fake science but using the way nature forms the Universe it is very possible.

If I was unable to find the very point from where the universe started I then could never be able to explain and prove the four cosmic pillars. These laws are the very pivotal fact around which science works.

Yea, now you think this is just another researcher shouting about something less important than yesterday's weather and then trying to become important by trying to make it sound as important as it never will be anyhow. Every discovery that science comes up with every time then gets blown up by the media to make it sound as the biggest hype that came this millennium. Every doctor doing study in this or that field say find that by beating yoghurt such a diet will increase your life by at least a thousand years and then make it sound as if this came from the Bible. Then later on one finds that it was a Yoghurt sponsored study and the yoghurt industry paid the University millions in donations. This is not the case and that I promise you. They use a lot of money to blow the reality and truth out of all proportions because it is not the research they sell but the advertising of yoghurt that has all the importance. Finding the point from where the Universe grew showed me the point whereto the Universe grows.

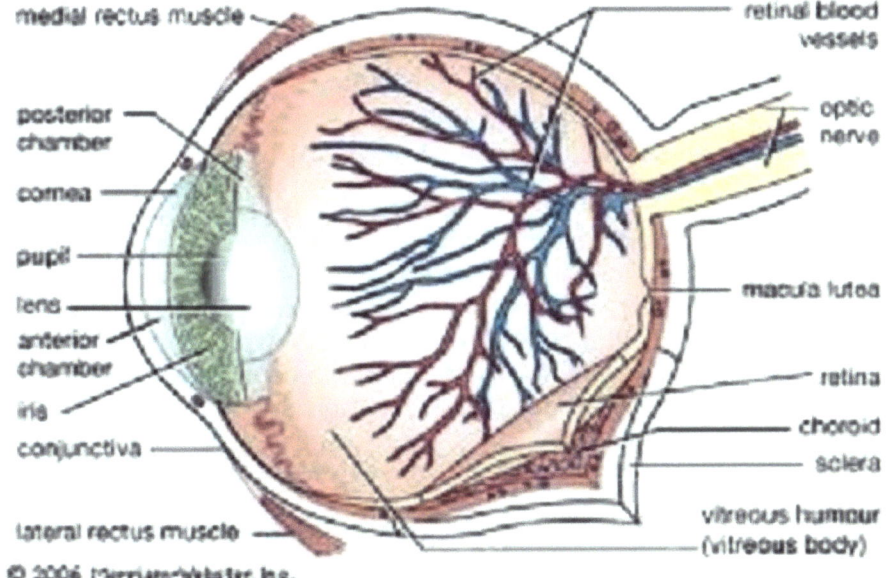

© 2006 Merriam-Webster Inc.

Without knowing where the Universe started I could never be able to determine how the Universe started but by locating where the point was from where the Universe started that knowledge enabled me to see how the Universe started and then developed. The process is far too enduring to explain in such a small confinement as this website offers but I will give you a hint and allow you to make your conclusions before purchasing the book or in the case of the Academic Introducing of the Titius Bode law then it is the books.

The answer is found within something so clear as your vision would allow. You can see because it is in front of your eyes but with your eyes you will never see the answer.

Understanding my concept is as easy or difficult as understanding how we see what we see. Can you think how we see a horse or a mountain? The horse or the mountain is many times the size of our eye but still we fit the horse or the mountain inside the eye. The answer has to do with your ability to see outer space as you see outer space as we witness the entire outer space that we are able to see. The answer is in the way you see outer space while you gaze at stars at nighttime. The centre of the Universe is in the picture that you see of outer space's vastness.

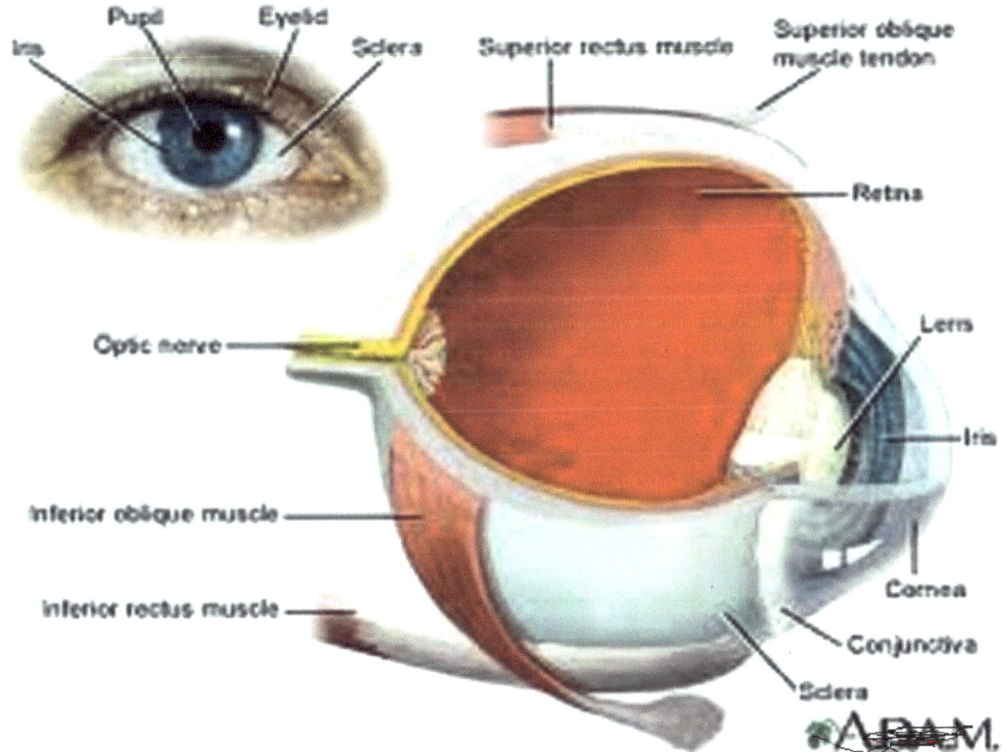

Have you ever sat back and gave it a thought how you are able to see all the space you are able to see at night…and if you think you can see it by light please do not read further because then you are just another one of the Newtonian simpletons that improvise their lack of intellect by reducing the complexity of the issue and to simplify is the same as compensating for the lack of clear understanding.

Go outside and look at the night sky. See the vastness you can see out there. There is no man that has the mathematical ability to measure the space that we are able to see at night. There is no means of expressing the volumetric measure of the space that we are able to see at night. One little speck becomes a galactica many times bigger than the Milky Way which we are part of. …And that speck is part of a vastness that goes beyond the understanding we mortal could ever have, except if we are Newtonian physicists…then we will simplify the issue to fit into our understanding instead of finding the true complexity that the situation offers. To belittle Creation is the Newtonian way of being smart.

To understand the vastness of the complexity of Creation take the conclusion down to the true dynamics. Try and see how by seeing you are able to fit that vastness of space holding more information than we have atoms on earth into the size of your eye. How can you locate all the space you are able to see as outer space and fit that sizable overwhelming information down to the size of your eye? How can the space reduce in volumetric size to be able to compress into a size that your eye is able to hold?

Forget about all the fancy Newtonian formulas with which they bullshit the entire human race. Forget them getting clever by explaining a mini – Black Hole or the stupidity they portrait when explaining the existence of time travel through a space whirl or finding Dark Matter that is going to pull the Universe back from expanding. Try to understand how the Universe will reduce so that it will fit into your eye and use that to locate the centre of the Universe. You know the size of your eye and you are unable to know the size of the visible Universe and then use that to determine where to find the centre of the Universe. The Newtonians always create a simplified Universe and then explain the Universe in relation to their small brainpower. **Now go and explain the Universe in the size it is and as large as it is and then arrive at true conclusions supporting reality. Your eye is small and the Universe is big.**

Meet the Newtonian physicist. IN THIS BOOK I try to introduce the reader to the brilliant Newtonian conspirator that has been dragging all of intelligent man by the nose for three centuries on a string. He promotes what can only be forces of witchcraft because he is unable to prove how the four forces apply the magic of gravitational pulling and yet he preaches the power that the forces generate. The Newtonians are not practising witchcraft but they are echoing a very famous alchemist that did believe in the power of magical forces and he created the magical forces of gravity as he invented the idea. The more the conspirator pretends to be an intellectual physicist the better a fool those conspirators become.

He   our brave and dapper Newtonian looks sheepish because he acts sheepish because
as   he follows he never questions what he believes and brainwash students to do the
                     same. Read how clever the physicists are in hiding their stupidity from students and the
                     public alike. He puts everything down to "black magical" dark matter that has no mass
and   yet will by "mass is pulling mass" save our Universe. This is because to him gravity is a
force   of magical proportions" and that he believes for he knows no better. He proves nothing to
                     students but by enlisting thought control those teaching physics force students to believe
in   science by applying some cruel mind-bending manipulation and to have students
                     believe science that has magic powers he makes them accept the unexplained. If you
believe in their brilliant mathematical genius I advise you to brave your mind for a big surprise is looming… The truth is that entire cosmos formed and still increases relevancies because of four cosmic pillars and science doesn't even know this or are able to explain why and how these phenomena forms or are in place. These phenomena prove that there is no "mass" pulling anything because of gravity.

I have asked this many times before but I ask again: Have you, the person reading this, ever thought how it is possible to see that much information that you see at night when looking at the sky with only your eyes? Ever thought about how you are able to see when you see everything in the night sky and how that much light information can fit into such a small space as your eye? Have you ever sat back and think what vastness in information it is that you see when you see the entirety of the Universe when looking at the Universe at night and what the size is of everything of that which you are able to see?

Those physicists formulating science are a mafia-gangster club controlling the dishonesty that formulates science… I am one person trying to correct what is a joke but what is also sold as the truth. If you wish to silence me, prove me wrong and then I challenge you to prove Newton correct or then accept the blame!

Science force humanity to accept the hoax Newton founded and science brainwash everybody to disbelieve nature while clinging on Newtonian hogwash. That is mind-control and that is brainwashing!

I have sent this book with six other books to eighty-five publishers and e-mailed this book to thirty something more. I had no response…not from one. This book opens a new era in understanding how the cosmos works…and not one publisher found it interesting enough to publish or to reply reasons why they found no interest in this as a publishing project. No one is prepared to break the hoax and publish the truth… You can go against God and you can criticize religion…any religion but do not show that Newton went wrong or had flaws…that is inexcusable.

For the first time in all of human history there is a method deciphered to show how NATURE no less forms the solar system…and in eighty five DVD's sent plus another (about) thirty six or seven e-mails going via sendspace and not one was interested to publish what I sent them...and still you don't see a conspiracy. Go to the Internet and see it is said this code can't be deciphered but I did find a way to decipher. Deciphering the Titius Bode law mathematically science says is impossible and yet you will read how I did it... and it is simple! Science plainly ignores nature while nature is the only reality we have.

Nature is the only reality but science brushes nature off the table, as if nature is madness. To so many publishers I sent the entire book among which 75% were Universities…I sent it as a unit with two chapters more than the book you read and found no publisher prepared to take on science and correct the hoax Newtonian science is.

Now you can find out how to crack the code by which nature (not Newton's fiction) forms the Universe in the manner that it forms the solar system. It is simple; it is adding 3 plus 4 to get 7! Finding the Titius Bode law is 7 also turns what you thought was cosmology into an explanation of the truth while the truth turns cosmology into truthful science. This Titius Bode law, its 7 / 10 or 10/7.

Should you feel I am unnecessarily hateful and negative in my attitude to Newtonian physicists yes I agree but I did not start the insults, they did? I never wanted to take on anybody but merely was on this quest to find what it is that I was unable to understand Newton. It was an honest search on my part to discover what there were that I was missing.

Then for years I had to endure the insults that I was not gifted enough to understand Newton or that my vision was blurry because if only I had a BsC degree in science or that I just was more informed about science there was a chance that I could understand Newton but seeing what my present status was I should just accept my fait and live my life according to my mental disabilities.

Then one day I went all the way to the University of Potchefstroom where I had an appointment with the senior professor there. In went and showed him a book that I wrote and the results I concluded. I am not going to call the man by name because mercifully I can honestly say I can't remember what his name was. The ma really got insulting and told me that if I only spent more time in front of genuine science books and less resorting to my imagination I could understand Newton. He said it was not his job to try and show mw what everybody on earth so clearly could see and I was in a fight with my underdeveloped mind and I then reflected my inabilities onto science and moreover onto Newton.

He then said I am a nuisance and I should keep my self to what I know and stop trying to reach what was clearly above my abilities. It came down to the point that with my feeble understanding there was no way any person could assist me in what I was incapable to grasp. I showed him that the Titius Bode law prevents planets to move to the sun thus the pulling of Newton was a completely corrupt idea. If the planets are distributed according to a proven formula since 1766 and there is no proven inclination that the set-up changes in accordance with the mass distribution of the planets then we must start to see that Newton is a spent idea and keeping Newton, as the main foundation is completely wrong.

Then I also drew his attention to the working principle of the pendulum. If the pendulum reads time and such time is in ratio not with the mass that the pendulum has but in accordance with the rotation movement the earth has, we must rather believe that it can't be mass that forms gravity but it is the movement of the earth's rotation by which we see gravity. If mass did produce gravity as Newton says it does then a long pendulum will read a different time than does a short pendulum arm?

Boy did this tick him off! Here comes an illiterate unschooled novice trying this to teach this Brain-box about thinking in terms of physics. He really laid it into me and the man was pissed!. He did what he could to belittle me and he went to task to set me in my place. He did not spare punches but neither did all the others I went to see. But they did not try to be as blatantly hurtful as our Potch professor. However he was just cruder than the others were but in the end they were all the same but for an aggressive tone in the voice. This then made me decide I am going to change my attitude from that moment on.

For many years I took the blame that I couldn't understand Newton but slowly it dawned on me that nobody was willing to explain Newton in a way that I could see my foolishness. They could blame me for not being educated, for not understanding but never could they show me what it was that I was unable to understand! Yet, in all the years the only evidence that they could present about what it is that they say they understand as proof about Newton is that they can fill a Universe with nothing to a point the nothing expands and becomes so much it spills over as the Universe grows from nothing becoming more. That is their understanding of Newton. They have filled a Universe with nothing while everything that the Universe is filed with they don't understand! When they were caught with their hands in the cocky jar as nature proved the Universe is expanding and not contracting as Newton's formulas insist, they created unseen Dark matter that will draw the Universe back into contraction because the Universe better start doing what Newton said it does!

After all my research I could find no evidence of Newton in nature but that was beside the point to those in science. What I found in nature those in science did not recognize! How can you discard what nature is in favour of Newton that is NOT in nature even in the slightest!

I then and there decided that all their blaming me and insulting my intellect only serve as proof that they had no answers for me. They could not prove Newton except just plainly accept Newton and if they can't prove Newton then I am correct in judging Newton is rubbish. I realised after many years of studying that the Universe came into place when 3 became, 4 and that turned to 5 on both sides of 2. That is why one has to add 3 and 4 and then divide by 10. This is all with the interaction of the Triangular law of Pythagoras. By moving 3 goes square to be 9 and four goes square to be 16, which forms 25 and that is the square of 5. This is how the Unversed started. It started when $3^2 + 4^2 = 5^2$. I am able to show how and where gravity starts and that is because I am unable to understand the bullshit they call Newton. This is how simple it is.

The axis forms by a line that holds three in place. This is singularity.

One more number and this is the circle or Π. This is the non-existing circle where space starts to apply form

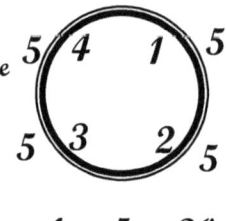

$4 \times 5 = 20$

Another number added and this is the numerical value of the circle or Π. This is where space ends and the cosmos starts

In this we find the value of how Π forms from being in singularity to forming material. Also in this we find the explanation of the Titius Bode law, the Lagrangian points, the Coanda effect as well as the Roche limit. From this point on I am able to explain gravity as it forms by Π. These laws are in the Universe at the present and are what is forming gravity. Newton's bullshit is not but now I must remain polite and show respect while they can insult me as they wish because they are clever and I am stupid...because they know everything and I know nothing! I decided I am going to give it as much as they can give it and let us see who in the end, the very bitter end even after 80 years will be the laughing stock and who will go down in history as correct. I am, not scared because nature backs me up while they only depend on their bloody stupidity. Some how in the future they will become the laughing stock and so I treat them accordingly.

I have studied and now I know how the 4 comic laws apply and how these laws form the cosmic code. I can explain what nature put in the Universe in place of all the fancy rubbish they create to hide everything that they don't know. Then why must I be quiet and pretend to be the lesser informed of the lot?

The Universe is what Kepler said it is. Kepler proves what he said years before Newton and that lot came into the picture. Let them silence me by proving Newton in any form, as Newton would apply in nature! Prove Newton and I will be the fool! But until then they are the biggest fools that ever lived!

The Universe form s by three divisions
Infinity (singularity)
Definity (materials)
Eternity (time as space)

Infinity, Definite and Eternity:
The only borders in the Universe

Infinity: The part of the Universe that can never start

Infinity: The part of the Universe that has no inside

Infinity: The part of the Universe that cannot move

Eternity: The part of the Universe that has no outside

Eternity: The part of the Universe that can never end

Eternity: The part of the Universe that cannot stop moving

Definite: Forms the part that holds movement and therefore defines space.

Definite: which is what name that I gave material

Definite: is the part that converts eternity into infinity or release heat back to eternity

Definite:

Definite: is the part in the Universe that that defines the Universe for what it is by keeping infinity apart from eternity.

Definite: keeps time in infinity apart from time in eternity

In this book I explain in detail the layout working process of Titius Bode law

The Titius Bode law a k a...
How the Solar System Forms

There is no other website going by the title of *How the Solar System Forms.* There are many that say *How the Solar System Formed* but this is the only one that reads specifically and explicitly *How the Solar System Forms.* This is because *How the Solar System Forms* totally annihilates their "**scientific idea**" of *How the Solar System Formed* because I go with nature and NOT Newton. In nature we find a law applying called the Titius Bode law. Before I get to the Titius Bode law I wish to relieve you of some of the brainwashing mainstream science is subduing everybody on earth with. I'm getting to the Titius Bode law but before that I wish to disprove what you have been told by mainstream science up to now.

There are four laws by which the solar system functions.

This is the Titius Bode law

The Titius Bode law proves that mass has no place in science. See in the picture how random mass is and with such randomness, how can mass place planets in the positions they hold? By my effort to solve the mystery of the Titius Bode Law, I prove that gravity forms not by mass but gravity forms by π forming in movement π². Solving the Titius Bode Law and proving from that how gravity works opens up a new view on the cosmos.

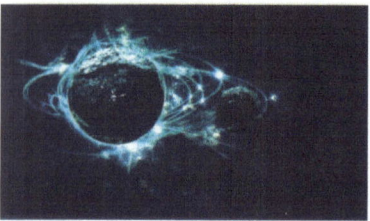

This is The Roche limit

The Roche limit has been around for centuries and with all the mathematical splendour available to apply in order to fathom concepts behind this phenomenon, still with all the computing ability of a machine all those physicists with all the mathematical superiority could not touch any understanding about the concept forming the background. Yet when using the truth about gravity in physics the answer is simple; it is that gravity is Π.

This is the Lagrangian points

The Lagrangian points have been known to science for centuries and with all the mathematical splendour available not one calculation could ever explain why this event is taking place. The satellites form precise locations positioned around the major planet and never comes closer while remaining in their positions.

I'll bet my shirt on the fact that most if not everybody never heard of these laws and yet they form the solar system as it applies gravity. Do you want to know why you are unfamiliar with these laws?

The Titius bode law is well known since it was discovered by Johan Titius in 1766 and was formulated as a mathematically expression by J.E Bode in 1778.

Since 1766 science knew Newton was a farce and since 1788 science had mathematical proof that this was a legitimate series of number positioning and yet we are 250 years on and still nobody in science made any effort to explain the Titius Bode law notwithstanding the apparent importance this law proves to be! Still it is mostly condemned as a "Fluke of nature" and has no purpose to investigate. Well when reading this page will change your opinion completely.

The Titius Bode law is how planets distribute places in the solar system. There is a precise sequence and order in which planets hold places and this annihilates the idea that mass plays any part in this sequence. In contrast the Newtonian system that science promotes currently says that gravity "pulls" according to mass. This is fictional. The mass of the planets is totally random and Newton holds no theoretical basis that nature supports. Look at the arrangement and you can visually see using your eyes that the distribution by mass is a hoax and is an invention Newton concocted and science validates for the past 300 years. Explaining the Titius Bode law for the first time ever breaks the 300-year-old myth and brings truth. Now we know how nature forms the smallest particles to build the cosmos.

The Roche limit is the law that says stars do not collide ever in spite of Newton's ridiculous idea that stars can or do collide. If two stars are in each other's atmosphere the law reads that when the minor of the two stars is closer than 2.4674 of the diameter of the major star, then the major star will liquefy the minor star into a gas plume that it then treats as more atmosphere. This is most significant in the cosmic principles I now put forward. There is not one instance or any evidence that shows where two stars do collide. When two or more stars are evenly matched by gravity the two stars do not collide either but become binary stars that spin around each other.

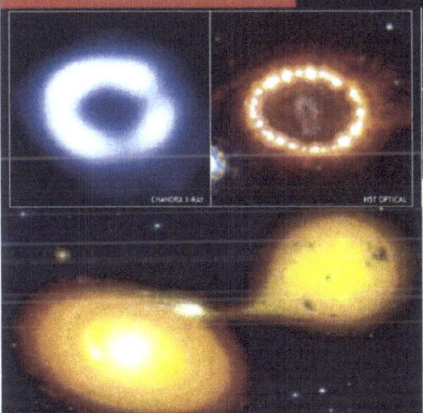

Stars or planets never collide and when a meteor enters the atmosphere like at Tunguska in Russia in 1908, the earth vaporises and liquefies the meteor and the meteor becomes fragments as well as more dust clouds in the atmosphere. This totally annihilates Newton's idea that a radius between structures diminishes by the gravitational attraction of mass. This book shows what happens in truth in nature.

The Coanda effect connects with the previous law. Gravity is a ratio that forms between what is solid space and what is liquid space. This applies when the movement of the solid (star or planet or atom) turns and by turning 7° it reduces the value of the circle of the space surrounding the solid from $\Pi = 21.991/7$ to the compressed value $\Pi = 3.142/1$ which connects to singularity. There is no pulling of material but only reducing of space when gravity contracts gas into liquid such as what the atmosphere is.

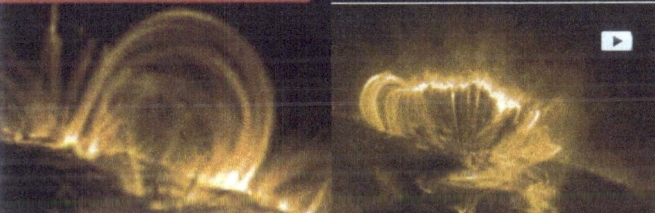

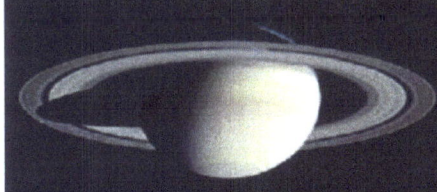

The Lagrangian Points This law proves that science is committed to deception for centuries. It is known that sattelites turn around planets in the same fashion as do planets orbit the sun. Saturn does not "pull" the satelites into its atmosphere or even "pull" the satellites closer and this knowleidge never drove science to question the validity of Newton ot his concept about "mass" pulling "mass". Even the rings form as debris spins in a circle that always remains constant and science should have rejected Newton's ideas on the grounds that nature never not once supports newton and Newtonian science in any way thinkable.

The Coanda effect

This is the interaction between different density levels in space. It is density differences that give us atmospheric layers and that is due to the Coanda effect interacting with the earth. Even the moon circling the earth is doing so on the basis that the Coanda effect provides. I prove that the atmosphere of the earth is the result of the Coanda effect applying and all about gravity interacting is the result of movement. Even the layers forming stars are the result of the Coanda effect applying differential density levels

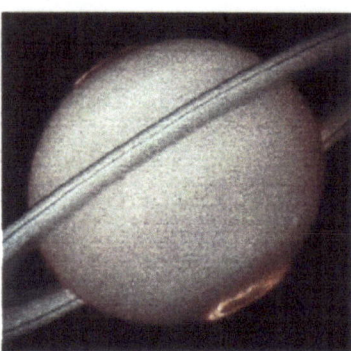

The Lagrangian Points

The Lagrangian Points system is the very law that disgraces all Newton's fables about "mass puling mass" and this law was known for centuries but also this principle was ignored for just as many centuries. This law proves that material circles structures.
The fact that there are structures or satellites circling around the gas planets are due to the Lagrangian points allowing the perfect positioning of orbits in a very controlled manner. There are always five structures set in a very specific manner as the structures align with the material that we call the planets.

The Titius Bode law

The Titius Bode law indicates a specific ratio that stars align with the sun. There is a specific line up formula where each star receives a position not in accordance with "mass" but with a location in the line-up of the planets. While this law is known for centuries it is blatantly ignored by science because this law drowns Newtonian thinking in muddy water. Because it destroys Newton' perception it was named as "a freak of Nature". This is what is in nature while Newton is nowhere.

The Roche limit and the Roche Lobe

When two binary stars enter each other's gravity fields they go into a tussle where each one attempts to destroy the other by exciting the movement of the other star. This only applies when the two duelling starts are about equal in size with no victor. This law also applies where the minor star is in the gravity range of the major star's field. When the minor star is within 2.4674 of the major star's diameter, the major star liquefies the minor star known as the Roche lobe and then dissolves the liquid the minor star forms into the atmosphere of the major star. This law proves again that starts will never collide but other laws in nature applies to prevent the "pulling" of stars

Have you ever heard of these laws? If you have I would be most surprised because these laws are not in any promotion or secured within cosmology's spotlight. They are in nature working as nature and nature has no other process of applying gravity than to form these four cosmic laws. Still science reject these laws and thereby reject nature in favour of Newton…and nature does not provide any inclination as to where Newton or his ideas could be found working.

Still all the while for decades I am rejected because I reject Newton and I reject Newton because nature rejects Newton. It is not I versus Newton but it is nature versus Newton…and you make the choice.

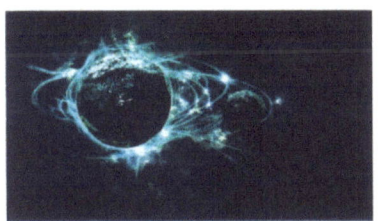

This is the Titius Bode law

The Titius Bode law proves that mass has no place in science. See in the picture how random mass is and with such randomness, how can mass place planets in the positions they hold? By my effort to solve the mystery of the Titius Bode Law, I prove that gravity forms not by mass but gravity forms by π forming in movement π². Solving the Titius Bode Law and proving from that how gravity works opens up a new view on the cosmos.

This is The Roche limit

The Roche limit has been around for centuries and with all the mathematical splendour available to apply in order to fathom concepts behind this phenomenon, still with all the computing ability of a machine all those physicists with all the mathematical superiority could not touch any understanding about the concept forming the background. Yet when using the truth about gravity in physics the answer is simple; it is that gravity is Π.

This is the Lagrangian points

The Lagrangian points have been known to science for centuries and with all the mathematical splendour available not one calculation could ever explain why this event is taking place. The satellites form precise locations positioned around the major planet and never comes closer while remaining in their positions.

This is the Coanda effect

The Coanda effect has powered turbine engines and aeroplanes in flight for almost a century and with all the mathematical splendour available to design the most terrific aircraft, not one engineer could mathematically compute one fact to show understanding why this takes place. How sad it is that those claiming of much superior intellect in physics remain just no more than having computing power. The understanding is not complex. I have to warn the readers that the topics are showing a very new approach with no quick answers. Understanding is in the proof and that does not come by reading just a few lines and then forming conclusions. The information is new but not hard to grasp. I did not put these phenomena in place and these phenomena nullifies Newton's correctness and the proof I bring goes beyond any doubt. I prove the Titius Bode law. Go to the internet and see how science doubt the Titius Bode Law and the correctness thereof while to solve the problem you add 3 plus 4 to get 7. That Is If you want to find a solution. I have published the Titius Bode Law in four already published books but in this one I go deeper than the four already published. In each of the books I present I disclose how the Titius Bode Law forms gravity. These books are:

All the while nature has this in place you have been brainwashed to believe Newton! Newton says that "mass" pulls "mass" and every person on earth are in mind control to believe this because the mind control is called teaching and schooling students. When you teach incorrect facts it is mind control. If I am wrong then nature is wrong and nature just can't be wrong! But science is wrong because Newton is wrong! ...And I base my facts on the way nature works and not on what Newton presumed. Your first reaction will be the urge to attack me but when you do attack me it is because you are subdued to believe what you were brainwashed from childhood to believe and believe in. You are going to NOT believe me because you are brainwashed to believe Newton's rubbish and rubbish it is.

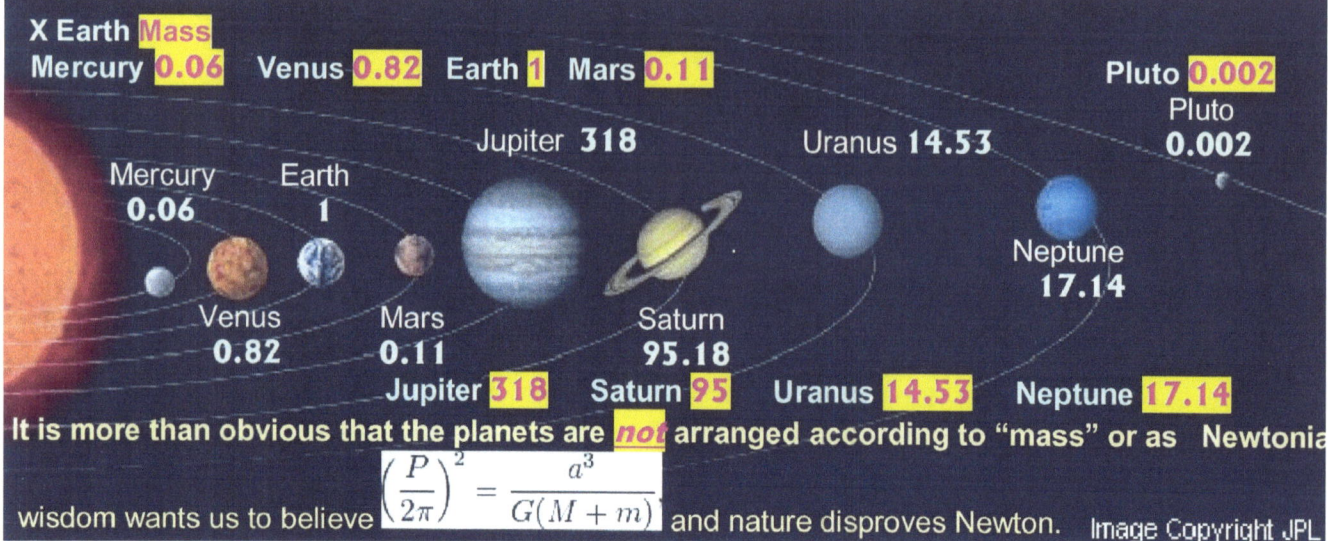

X Earth Mass
Mercury 0.06   Venus 0.82   Earth 1   Mars 0.11   Pluto 0.002
                                                                                   Pluto
                              Jupiter 318   Uranus 14.53   0.002
Mercury   Earth
0.06   1
Venus   Mars   Saturn   Neptune
0.82   0.11   95.18   17.14
       Jupiter 318   Saturn 95   Uranus 14.53   Neptune 17.14

It is more than obvious that the planets are *not* arranged according to "mass" or as Newtonia

wisdom wants us to believe $\left(\dfrac{P}{2\pi}\right)^2 = \dfrac{a^3}{G(M+m)}$ and nature disproves Newton.   Image Copyright JPL

Lets start the way that Newtonian surmise the solar system started. According to their Newtonian view there was this bowl of "gas" (and legal or not) I include pictures of how they say this process started.

Scientists believe that the solar system was formed when a cloud of gas and dust in space was disturbed, maybe by the explosion of a nearby star (called a supernova). This explosion made waves in space, which squeezed the cloud of gas and dust...and so how did that happen? Reading that it reminds me when I first as a boy of 5 read this crap AND I believed in it because I believed science and scientists.

Solar System Formation - Windows to the Universe
www.windows2universe.org/our_solar_system/formation.html

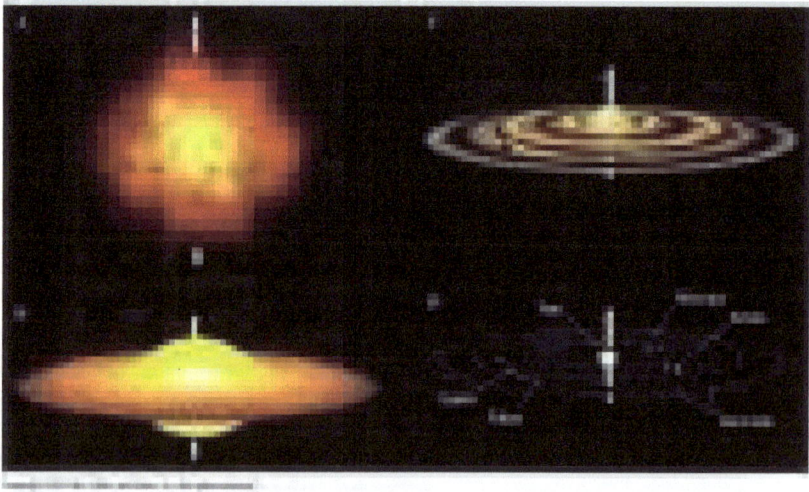

This is the accepted intellectual and official version explaining how the solar system formed:

To begin with it is notable that every (informed) website is about how the solar system formed and not one is about how the solar system forms. Why would that be? I'll tell you why: it is because they avoid this topic as much as they can because the way that the solar system forms denounces Newton totally. The Titius Bode law is discrediting Newtonians and Newton completely and so they call it "a freak of nature"! Every website you find is about this idea that the solar system came about from a cloud of dust that then by Newton's magical forces pulled and pulled and pulled and then the planets started. There is no date when this pulling started and there is no date when the pulling concluded to be planets eventually.

Now does this not remind of the three pigs and the wolf that blew and blew and blew and blew and blew their houses out existence? Oh sorry, they blew and the gravity pulled and that is the biggest technical difference there is. About the rest, well to prove the pigs are much easier and could be eventually much more accurate than the Newtonian myth of "mass" pulling "mass" anywhere and everywhere.

If you are a student ask them why did Jupiter end up that much bigger than all the rest. Why did it collect more dust? I explain this in some books where I explain the Titus Bode law principles applying.

This is the picture of circles that they use to boggle your mind and brainwash you into submission...and I challenge any one of them excluding not one of the Highly Educated masters to prove me wrong about brainwashing all students into believing that the circle sizes are as they portrait and the material sizes are in ratio just to cheat you.

These pictures are part of the conspiracy to brainwash students into believing their misconduct. This picture seems so honest but it is complete deception because realistically the following is the accurate portraying of material to ratio in the solar system.

The Newtonian idea is that the solar system formed as a cloud where material collected at one specific point from areas so vastly spread the distances goes beyond human understanding. From this one ball of dust material gathered at very specific intervals that at this moment goes by the name of the Titius Bode law. If it was possible one could fit all the planets forming the solar system into a line-up where the lot will have room inside the distance that there is between the earth and moon. This distance we find between the earth and the moon is large enough to host all the planets taking up space in the solar system. This picture gives no one any idea of what is applying in reality or is truthful to nature. There is the Newtonian impression and then reality is nature's version...Science stick with their funny make-believe version.

In light of this I am now going to describe and explain the layout of the planetary formation as it is in nature and as it is in reality. This is when I explain in detail what the Titius Bode law is. I will say this much: in ratio the planet Pluto is 100 times further than what the planet Mercury is from the sun.

The Sun
The distance between the Sun and Mercury
The distance between the Sun and Pluto

The Position that Mercury holds.   *The Position that Mercury holds.*

On the scale above the earth moon patrician will never be able to feature any space.

To give the readers some idea I will show the ratio applying between the sun and Pluto.

For those unfamiliar with the dimensions of a Rugby fields I give an example to show the applying distances. A rugby field is 100 meters from dead line to dead line and 100 yards from goal post to goal post. Half the field is 50 yards or 50 meters depending on from where the lines are you measure from.

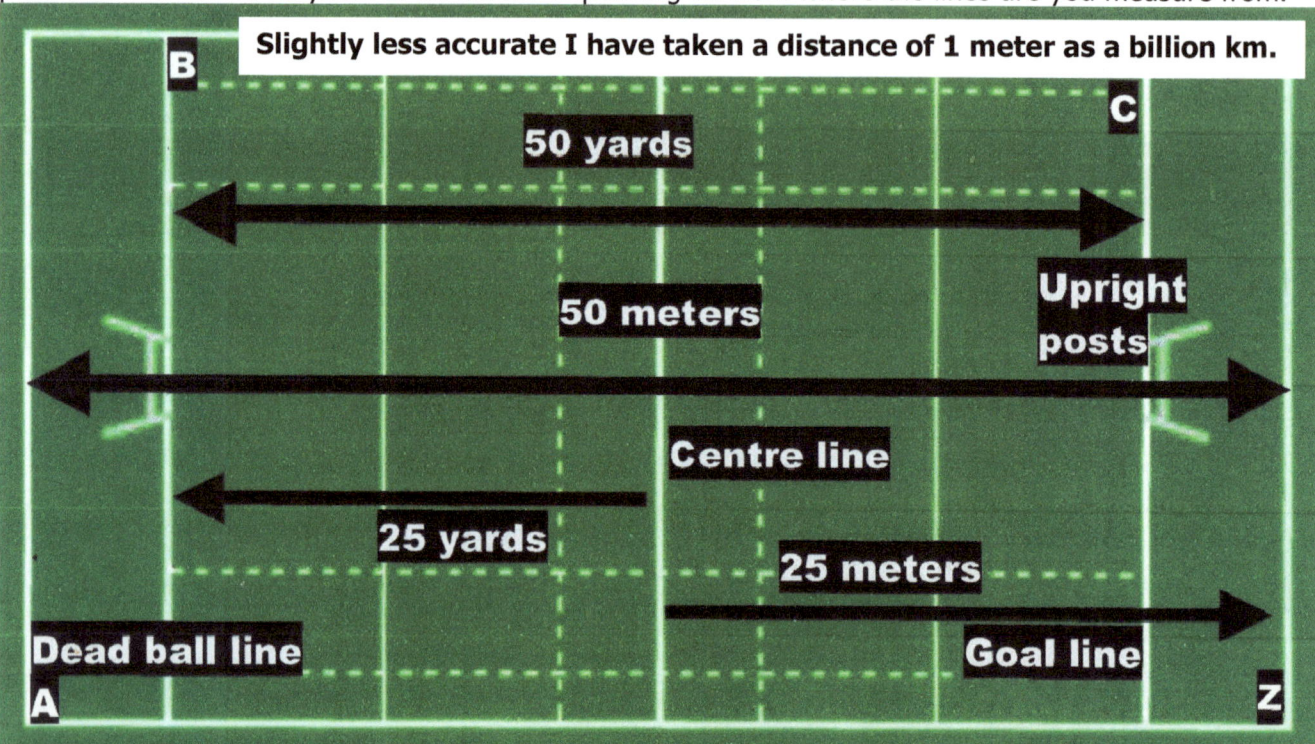

The following sketches say everything I wish to say to indicate distances and planet sizes applying.

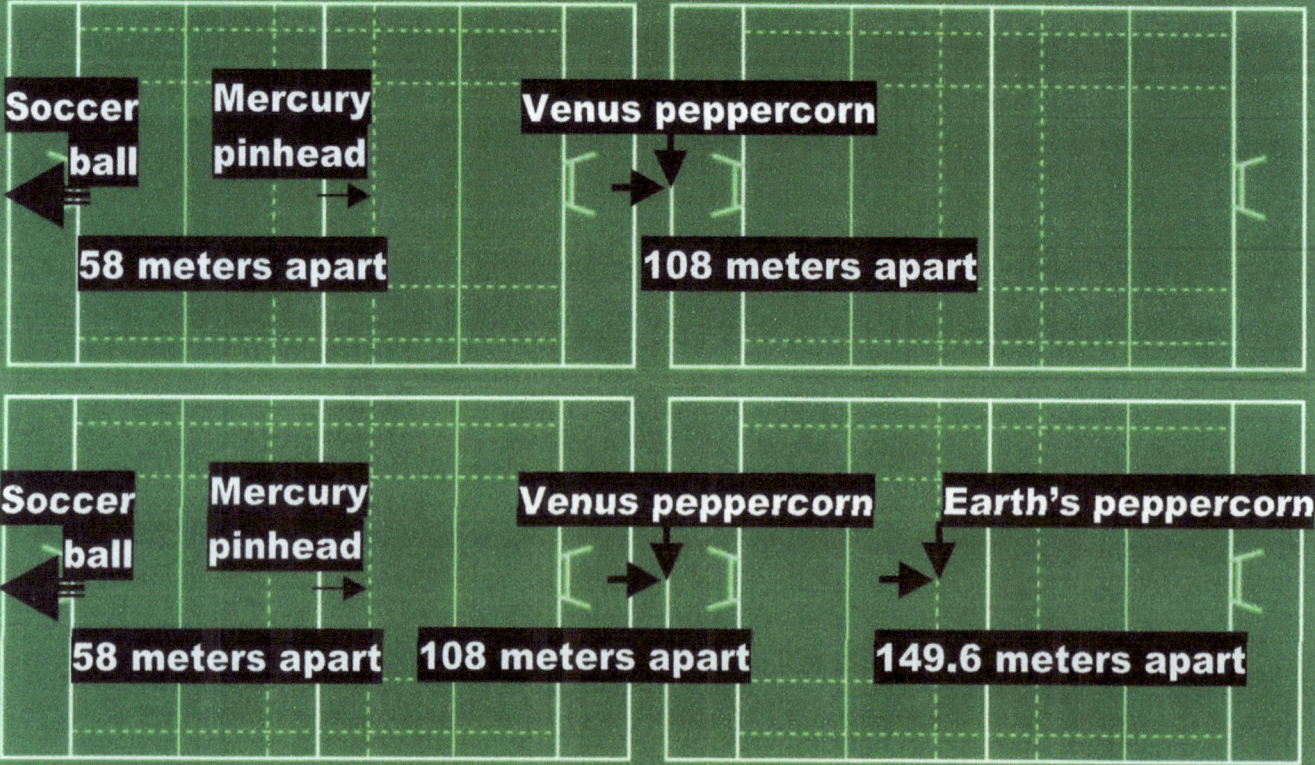

However, reaching Saturn I run out of space. Saturn on this scale even is 1400 meters from the sun.

100   200   300   400   500   600   700   800   900 1000 1100 1200 1300 1400

Soccer ball as the sun   **Saturn as a hazelnut**

In ratio Uranus doubles in ratio and then you have Pluto also almost doubling in ratio. To put Pluto's ratio in comparison I found senseless because you sit with nearly sixty rugby fields, a sun the size of a soccer ball and six kilometres further there is a pinhead that no one will ever see as these ratios apply. All the dust in this dome collected at one point so small in ratio it forms a pinhead...you still believe Newton?

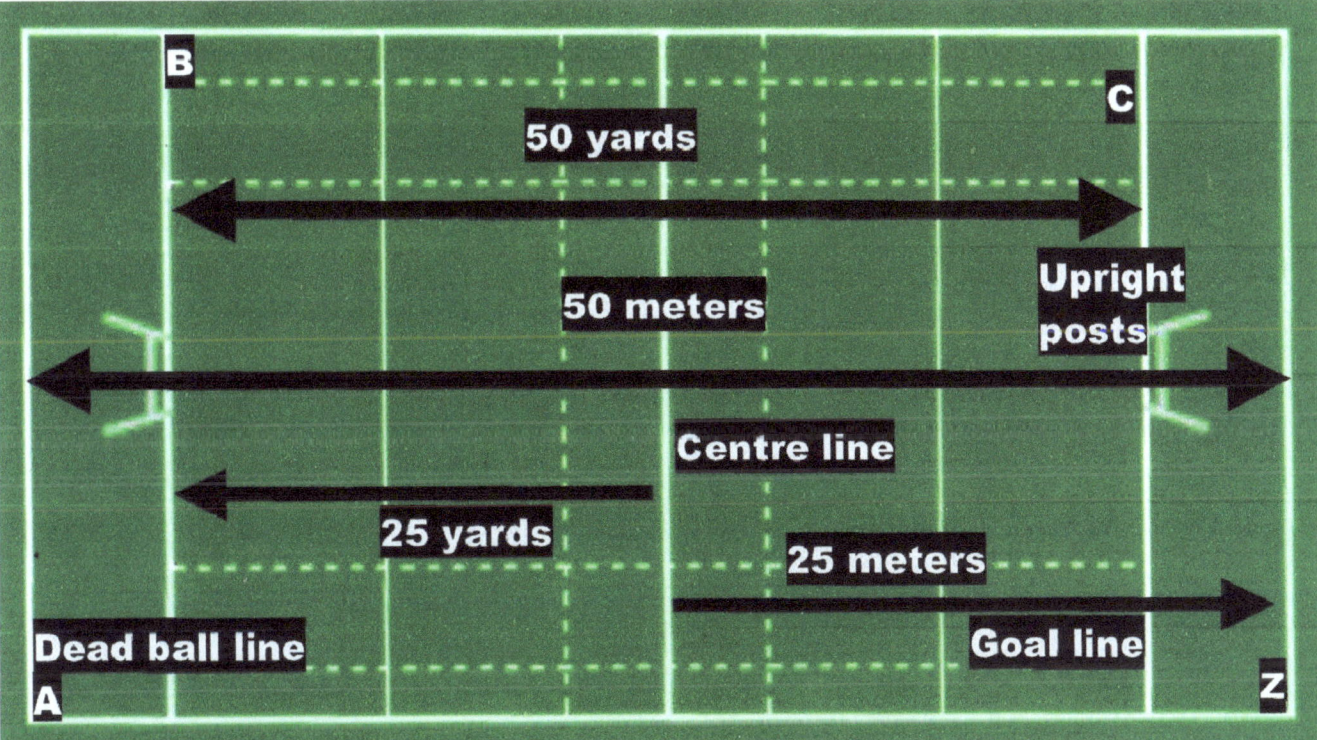

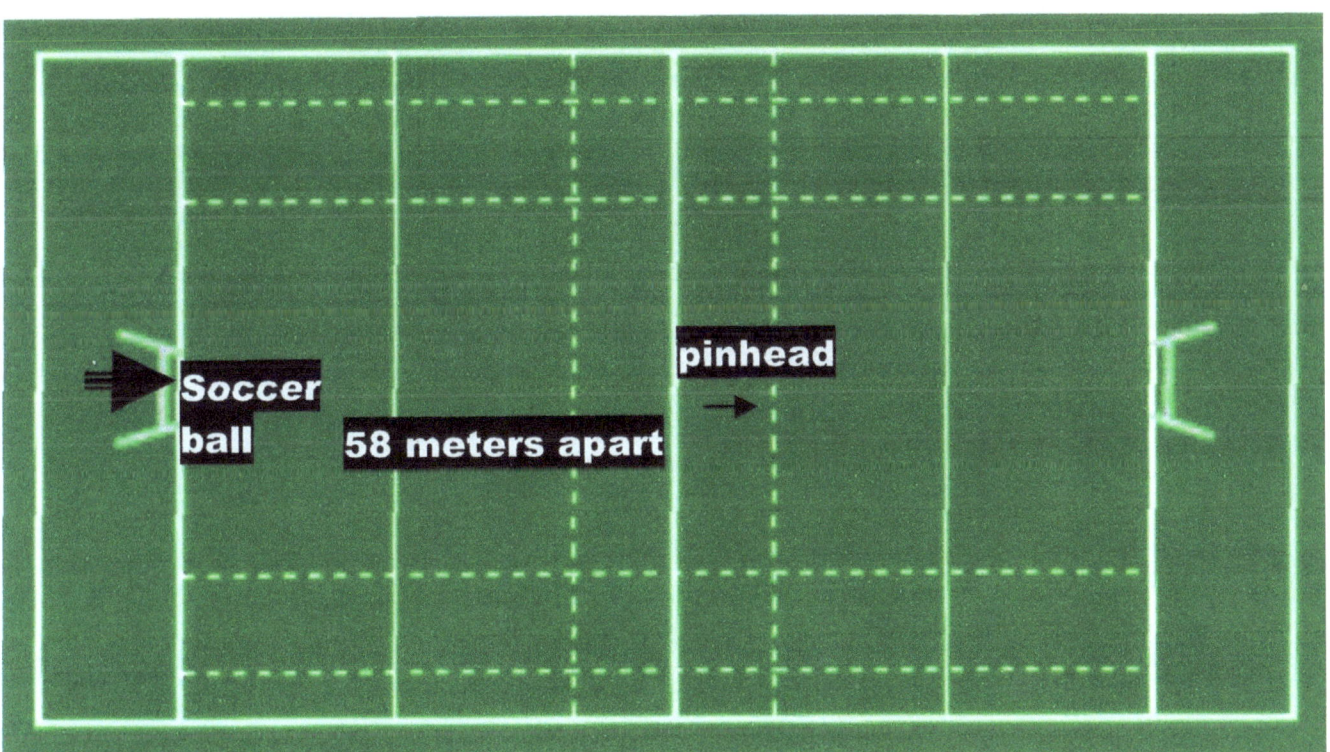

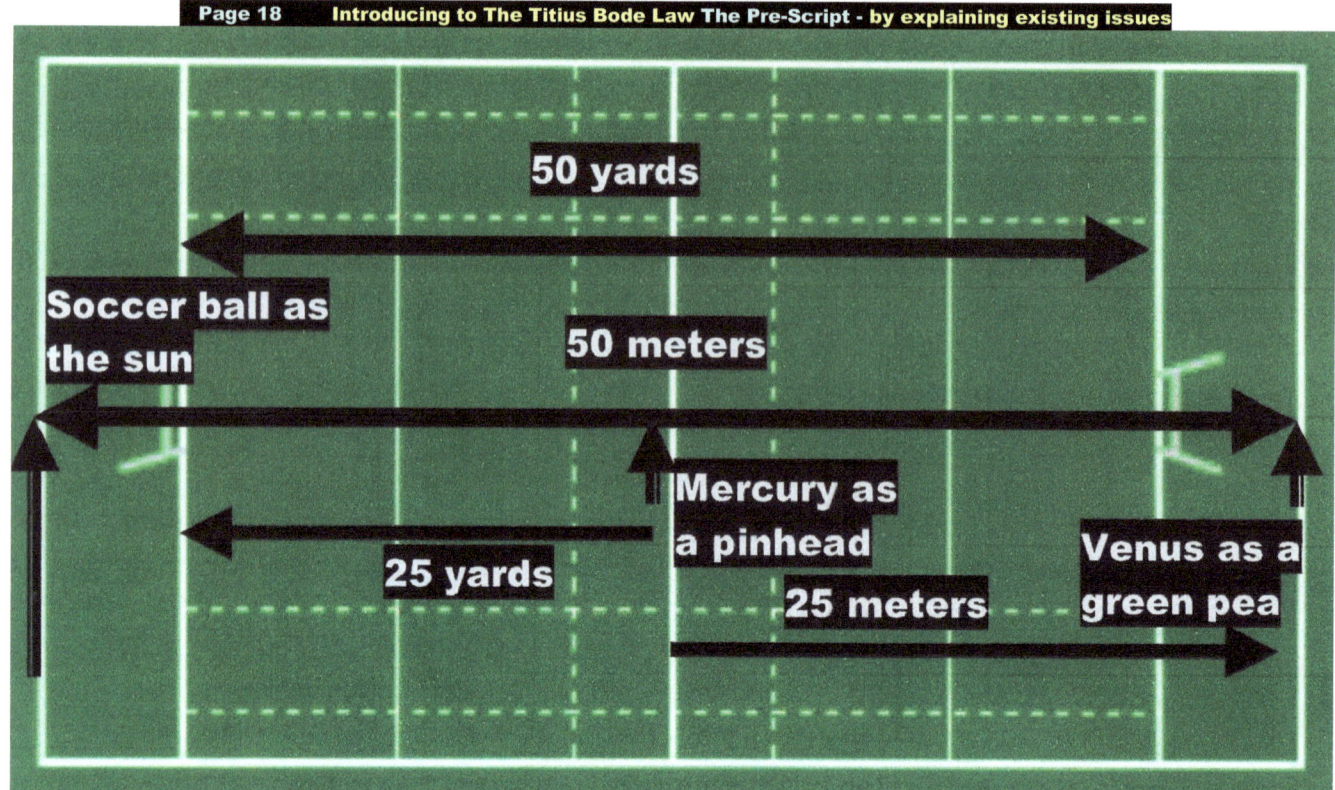

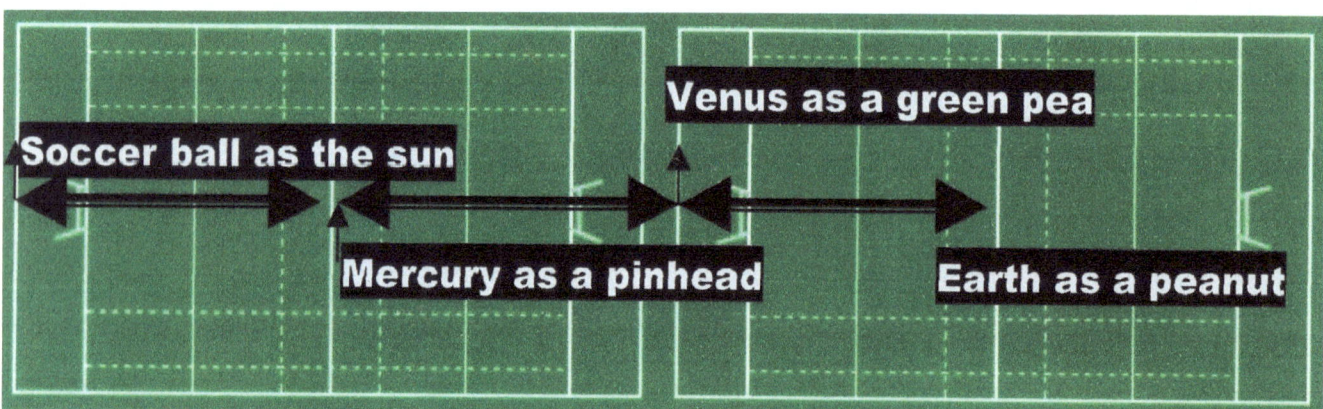

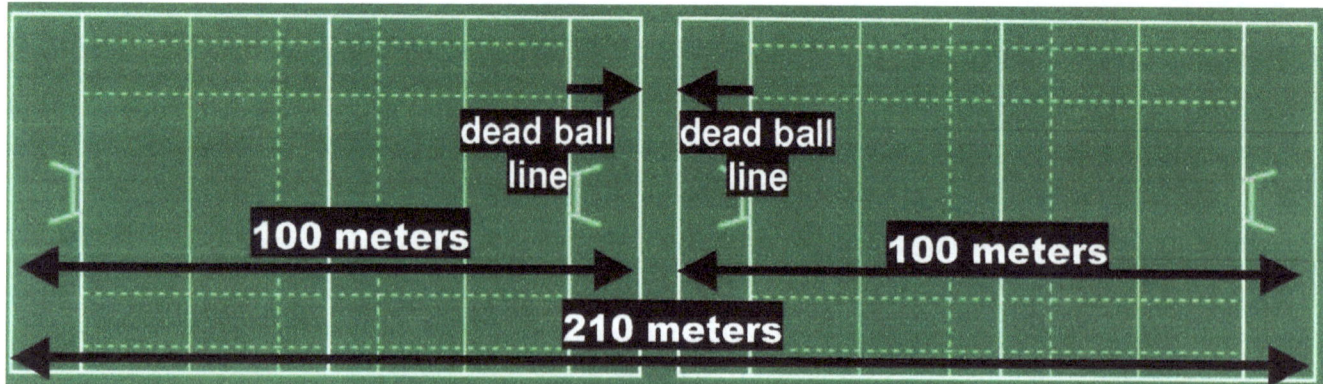

I am not going to explain this because it will take up too many pages. However study the pictures and it will become clear how the ratio works. In the books I go into lots and lots of detail concerning the workings of the four laws and especially the Titus Bode law.

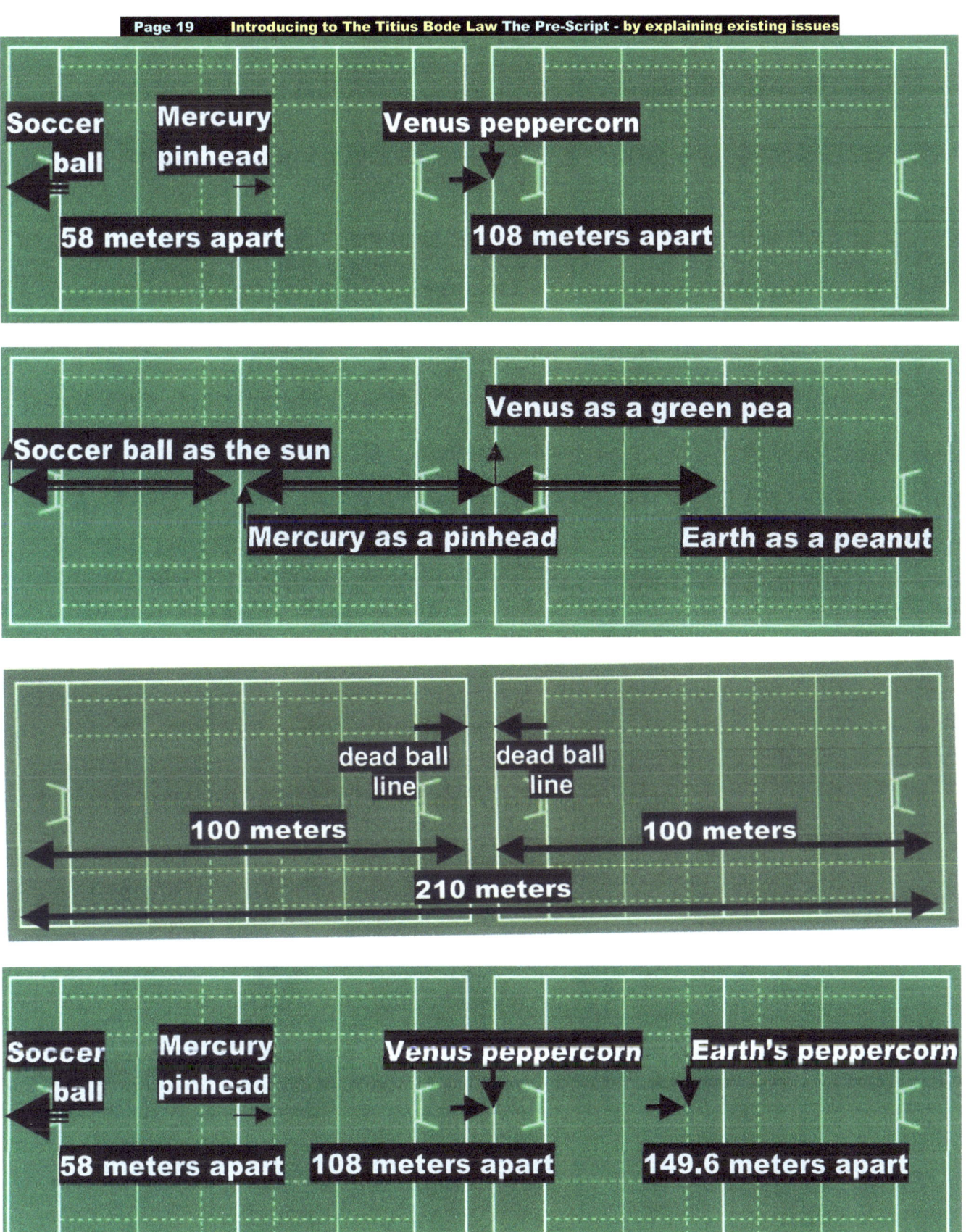

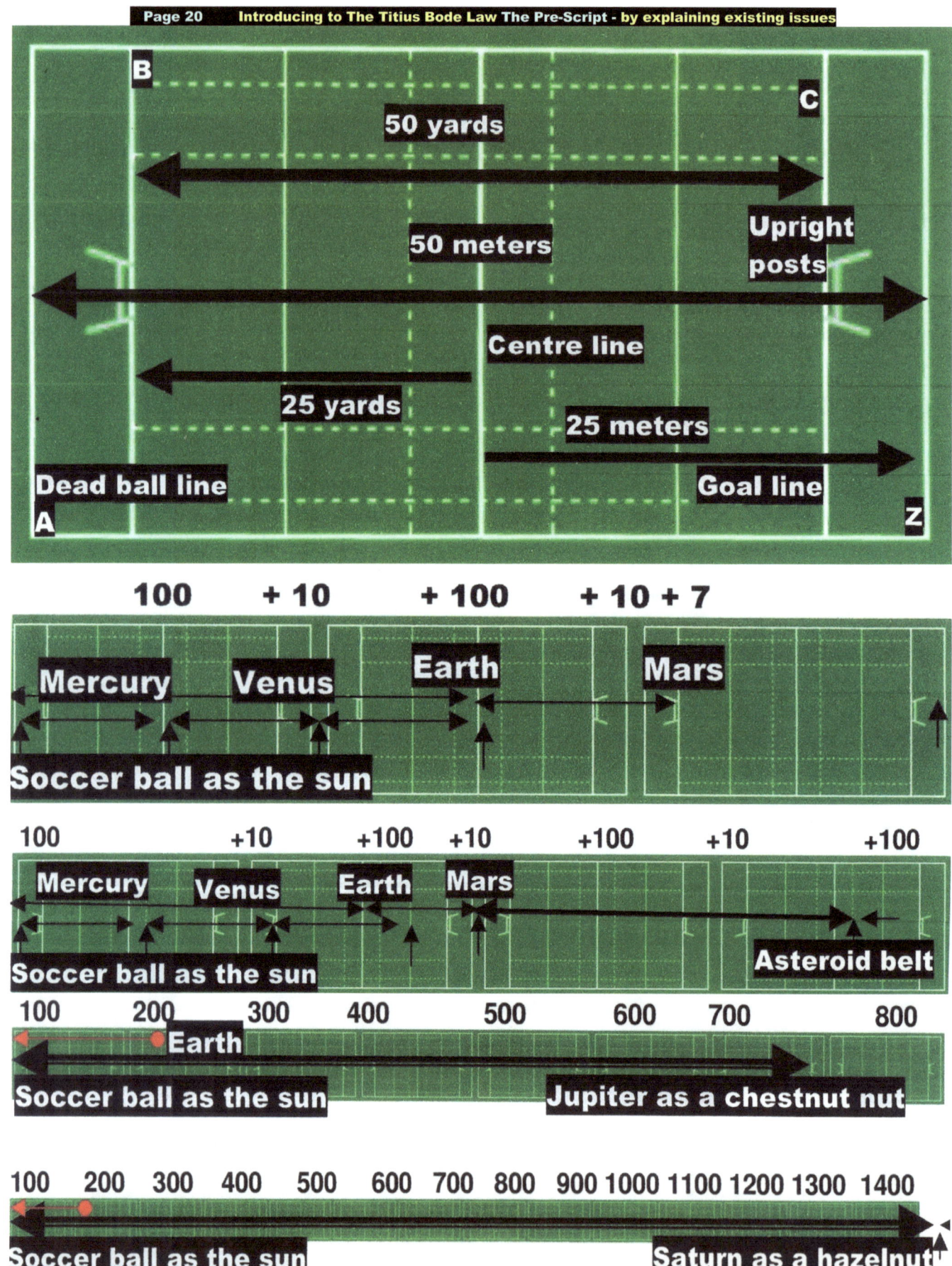

The big wake-up call comes when viewing the wide spread space increase with every planet orbiting the sun. The density change we have with every orbit places the planet in a completely new gravitational concept and this puts travelling even at the speed of light relevant to the density applying in such an orbit. This changes the cosmos completely and it is advised you read what is said before discounting the theory

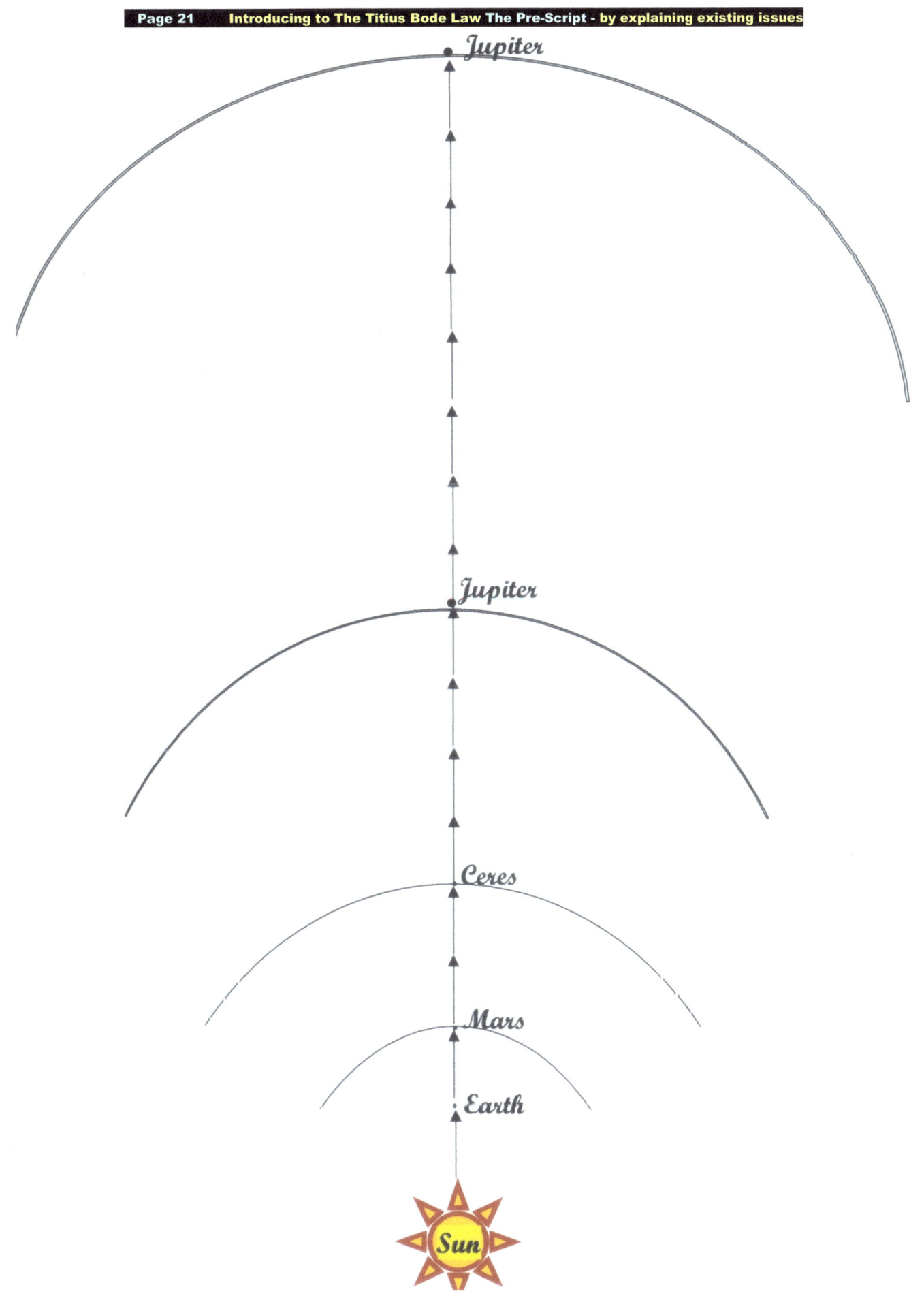

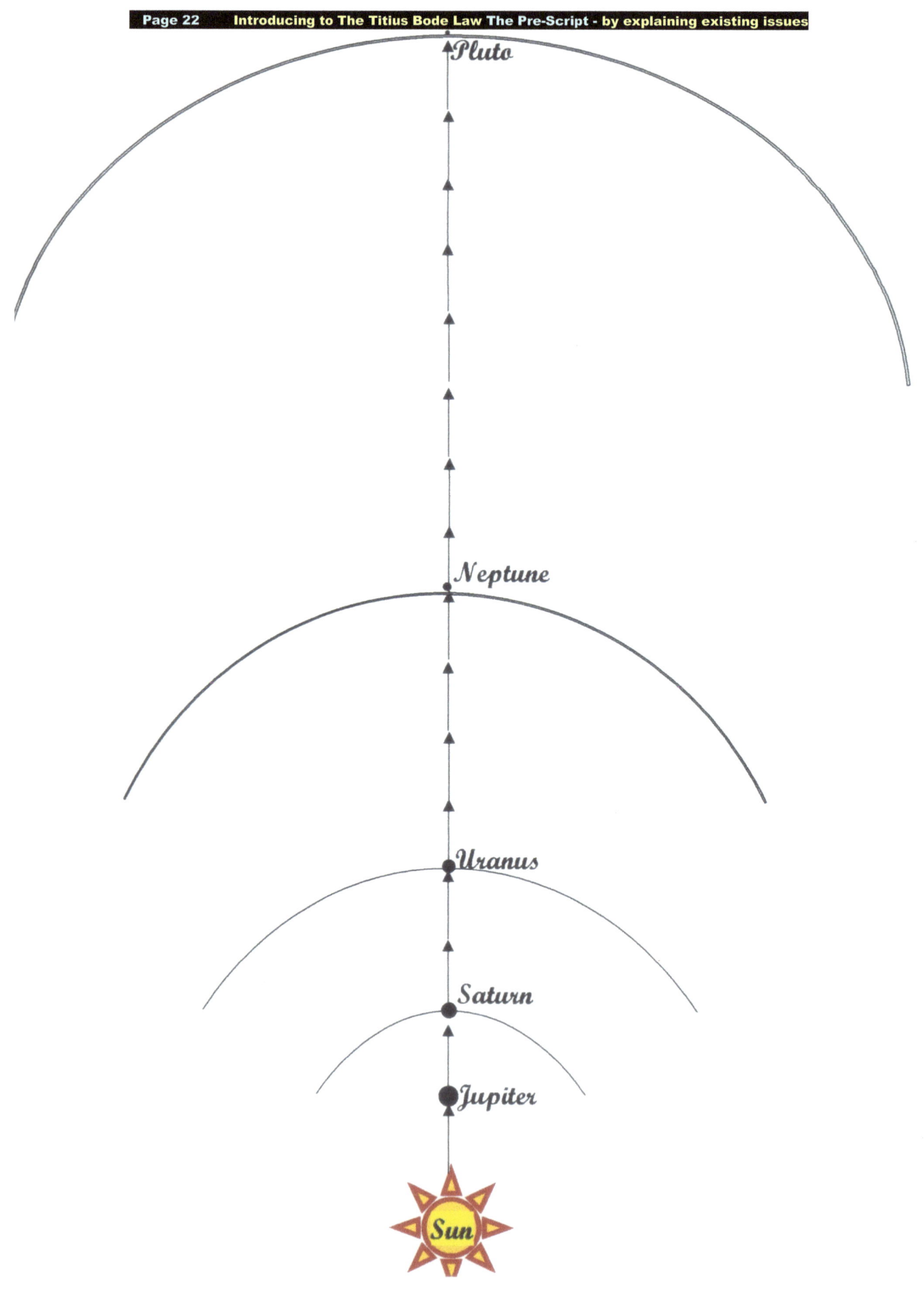

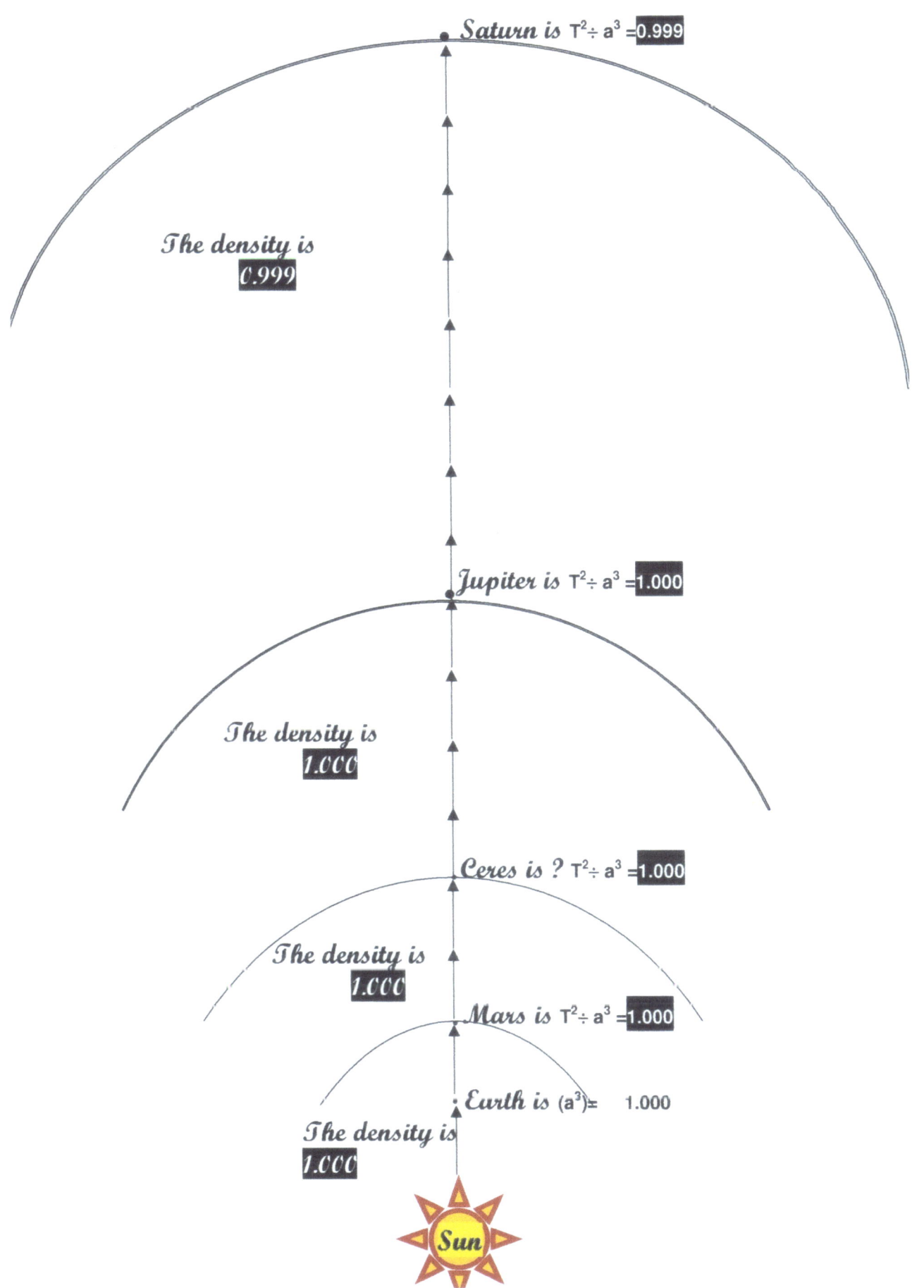

Saturn is $T^2 \div a^3 = 0.999$

The density is *0.999*

Jupiter is $T^2 \div a^3 = 1.000$

The density is *1.000*

Ceres is ? $T^2 \div a^3 = 1.000$

The density is *1.000*

Mars is $T^2 \div a^3 = 1.000$

Earth is $(a^3) =$ 1.000

The density is *1.000*

Sun

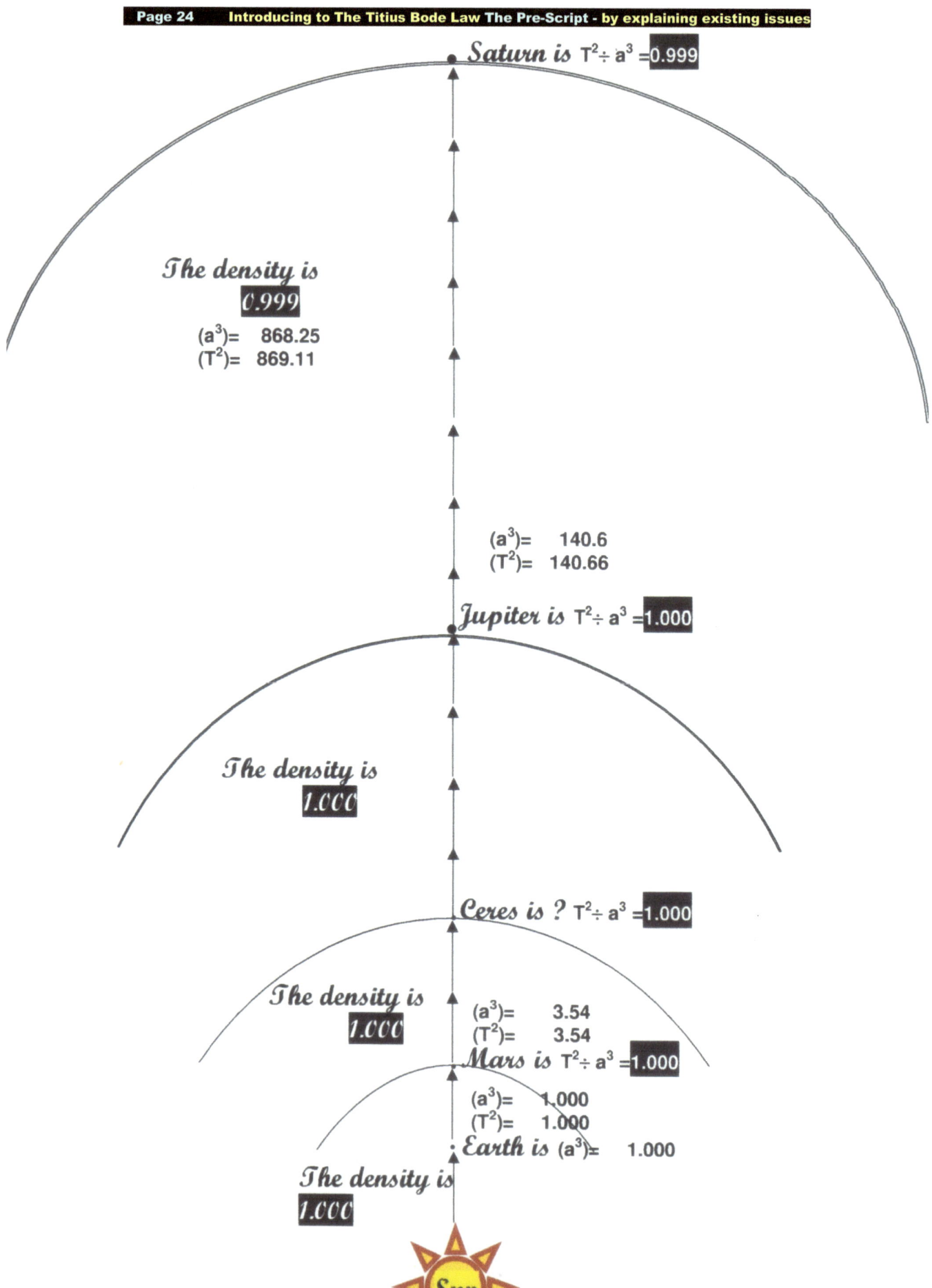

Saturn is $T^2 \div a^3 = $ 0.999

The density is 0.999

$(a^3) = $ 868.25
$(T^2) = $ 869.11

$(a^3) = $ 140.6
$(T^2) = $ 140.66

Jupiter is $T^2 \div a^3 = $ 1.000

The density is 1.000

Ceres is ? $T^2 \div a^3 = $ 1.000

The density is 1.000

$(a^3) = $ 3.54
$(T^2) = $ 3.54

Mars is $T^2 \div a^3 = $ 1.000

$(a^3) = $ 1.000
$(T^2) = $ 1.000

Earth is $(a^3) = $ 1.000

The density is 1.000

Sun

100   200   300   400   500   600   700   800

● Earth

Soccer ball as the sun

Jupiter as a chestnut nut

The size of material within the solar system allows you to fit all the space that the planetary material holds in the space we have between the earth and the moon. That is not much space and yet it holds all the material excluding the sun. Now the Big question to answer is how did a lot of dust accumulate all the dust into these small circles, which are the rotational circles that the planets have to circle around each one's axis.

Give me the formula that could be used and that can apply to calculate the force each one had to gather from the "dome" that the dust formed that grouped the particles into the units we now identify as planets. If they come up with

$$F = G\frac{M_1 M_2}{r^2}, \quad \left(\frac{P}{2\pi}\right)^2 = \left(\frac{a^2\sqrt{1-\in^2}}{\ell}\right)^2 = \frac{a^4\left(1-\in^2\right)}{\ell^2} = \frac{a^4\left(1-\in^2\right)}{a\left(1-\in^2\right)GM} = \frac{a^3}{GM},$$

$$P = \left(\frac{4\pi^2 a^3}{G(M+m)}\right)^{0.5} \quad P = \left(\frac{4\pi^2 a^3}{G(M+m)}\right)^{0.5} \quad \text{or} \quad \left(\frac{P}{2\pi}\right)^2 = \left(\frac{a^2\sqrt{1-\varepsilon^2}}{\ell}\right)^2 = \frac{a^4(1-\varepsilon^2)}{\ell^2} = \frac{a^4(1-\varepsilon^2)}{a(1-\varepsilon^2)GM} = \frac{a^3}{GM}$$

the whole lot is as fake as a three-dollar bill painted pink. Not one of these Newtonian formulas has any credence or truth and I wish students would challenge those educated professors to prove the credibility of these formulas. Every person that practice mainstream physics is guilty of hiding this conspiracy.

Earth   Al the other   planets   Moon

If you have the opportunity or if you personally are one of the few Super-Educated-Masters-of-Magic that can **Understand Newton** then please inform me about the formulas you lot used to calculate how the dust cane to be Planets and how long that process took. Remember you only have the Newtonian time figure of 4.5×10^9 earth cycles around the sun to complete and to show how every tiny dust particle pulled another tiny dust particle closer by the magic of using Newton's force/s. remember also to be very explicit about the distances that the dust had to cover to reach the very point where the planets at present rotate. Please also show how the "small" planets could muster the "gravity" to form such dense lumps and how the "big" planets only was able to get to the "gas" structure and were unable to compress further to reach the solidity that the "small" planets could achieve. This was forever one of the riddles I never could understand in all the time I could *never understand Newton!* Should you be able to *understand Newton* then sorry, but I have bad news; you are brainwashed by your superiors and as fatal as it sounds but your understanding is mentally completely disrupted in the past.

Then also how did the "mass" pulling process never got that safelights closer to he "big" planets or the moons to collide with their planets. What made the "mass" pulling fail; that stopped the moons and satellites to unite with the major structures that they orbit.

Stephen Hawking - Formation of the Solar System - YouTube

I challenge Professor Doctor Stephen Hawking, the most brilliant brain on earth to prove the total crap he comes up with! If I could use the baloney what he says I would never need any fertilizer or manure in my fields on a farm because the bullshit he releases into the world is so much crap it is toxic. It can kill because with that crap his killing all the young minds of students by brainwashing those minds with a complete and unproven fairy tale. I challenge him and any Great Mind such as he supposedly has to prove the process they advocate as truth. I say this because if he had a great mind as everyone says he has then why did he never realise that Newton and his ideas are completely fake. If there is any person that reads this and can get into contact with Professor Doctor Stephen Hawking or any person that is to his likeliness in cleverness give Professor Doctor Stephen Hawking or any person such as he is my e-mail address. I repeat my e-mail address **peet@naturescosmicconcept.co.za** **or use another one that will directly come to me:** mail.naturescosmicconcept.co.za

If you think I try to disprove Newton and mainstream science, then no I can't be credited for that. If I try to take that honour I might get sewed because it is some one much more important that has to claim that fact. The earth and the moon are moving apart at a rate of 3.78 cm or 1.48 inches per year.

In spite of what they teach you at school but all that baloney about "mass" pulling "mass", the Universe is telling a different storey. Nature says it's a fact that the Earth and the moon are parting further from each other.

Earth   *Moon*

This is what nature shows and you can check up on the Internet whether I am wrong. Can you see why I say they brainwash students into believing what is wrong. If the earth and the moon are growing that much apart, then think how far is the sun and Pluto parting per year.

Inside my books I show a shipload of more discrepancies Newtonian science try to hide. That is why I have written many books on the conspiracy in science to hide Newton's blunders and science has been doing that since **1705** when Edmund Halley supposedly used Isaac Newton's formula to plot the route that Halley's comet will follow on its way to the sun. At that point science went corrupt (if it ever was honest) because I challenge those in office to show how Halley applied this myth. The mass of the comet was not known. The mass of the earth or Jupiter was not known. At the time it was not even a global accepted fact that the earth is round. Only much later did science conform to that idea when experiments confirmed this as a fact! How did they apply the formula if they did not know the distances or the mass that applied in each case? This is a fib if ever there was a hoax in science. This is the biggest hoax ever. Can you believe that the greatest minds since **1705** were all fooled by this idea and nobody noticed?

I say Newton is wrong…why would I say Newton is wrong?

A meteorologist said on TV that hail of .5 inches would fall at 60 km p h. while hail of .75 inches will fall at 70 km per h. and then hail being 2 inches will fall at 90 km p h. This is a statement of a person that holds a doctoral degree in science and no less than being a meteorologist. If a master such as he goes that far off target when reasoning about gravitational physics how can a novice ever hope to understand physics?

This is the typical presumption going around in every person's head. Heavier objects fall faster and smaller objects fall slower. That is Newton and that is rubbish. This is a highly educated scientist with a doctoral degree in meteorology. Say we have a truck a petite dancer and a normal frog falling down from an aircraft. Then say the three were about the same size. We now know they will fall about at an equal pace. However that is in Newton's Universe where nothing makes sense because they fill the entire Universe with nothing and to Newtonians that makes sense!

In the total fictional world of Newtonians they give the nothing that fills the Universe a mass value and then they are satisfied with such rubbish. We have Galileo declaring all things fall equal notwithstanding "mass" or size. These are two completely apposing views and yet to the Newtonian science these two declarations say precisely the same thing and that is absurd. If the truck weighed 45 tons and the dancer weighed 45 tons and the frog also weighed 45 tons then Newton was correct and Galileo was correct.

However this scenario is valid in Newton's and the Newtonians' Universe but in reality and in Galileo's Universe all things fall equal and at the same pace notwithstanding and irrespective of weight size or structure. Now lets leave Newton's world of make-believe and enter Galileo's world of reality. Galileo said all things fall equal. This is not a statement but it is a fact and moreover it is what we find applying in nature.

There is the "mass" of the earth. You show me one formula in physics that ever apply the mass of the earth in any equation. The earth never has a usable "mass". It is always only the "mass" of the object that is used in any and in every equation in physics.

When the earth "pulls" an object whether the object is on the earth or coming towards the earth, the earth holds a "mass"

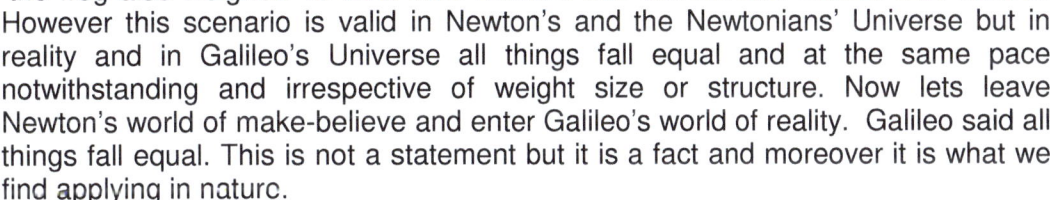

value of one. Never does the earth's "mass" feature in the equation. Then the earths "mass" has nothing to do with gravity and "mass" does not "pull" anything anywhere! Go make sense of that even if you are a Newtonian!

When this truck and this girl and this frog fall down to earth they do not to fall equal according to Newton that has "mass" "pulling" but in reality they do fall equal under equal conditions. That proves one correct to Galileo and nothing going the way of Newton. If Newton's presumption had any validation the truck must land hours before the girl with the frog not landing close to the other two because it took the frog a number of days to reach the earth. The truck has a mass many times over the mass of the girl and therefore must fall at a blistering speed while the frog is standing still waiting for the earth to turn so that it could travel to another continent and settle there as an immigrant. However we know the girl and the frog can get into and out of the truck as they please while falling at the same rate as the other falling objects.

Therefore when keeping these facts in mind then please if you can prove me wrong in this assessment then drop the book and never touch it again but if you are unable to prove me wrong or if you are unable to prove Newton correct then don't go and dismiss me because your culture tells you Newton is correct and no one should prove Newton as correct because after all we all believe Newton is correct.

Then we get to the comet. This is supposed to be Newton's speciality.

Newton left us with this formula $F = G\dfrac{M_1M_2}{r^2}$ and since Newton's departure everybody in science clung to this formula as if it was religion. I maintain it is just a science religiosity void of reality.

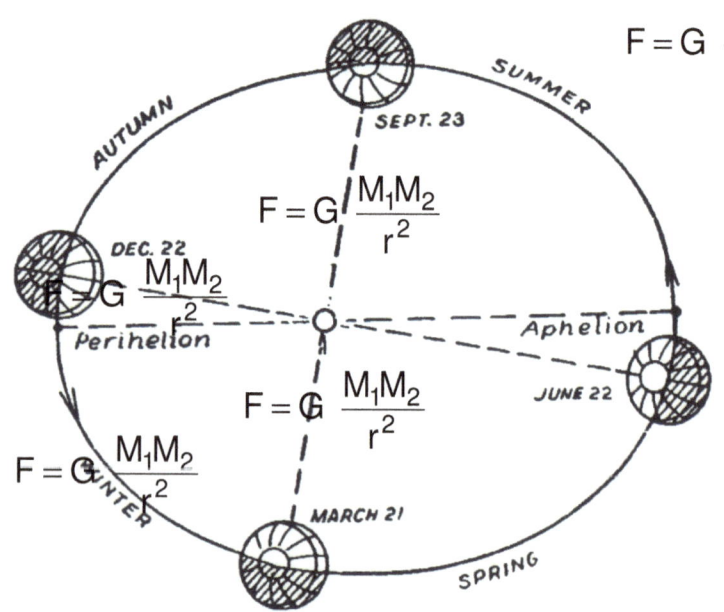

$F = G\dfrac{M_1M_2}{r^2}$ This is one of the disputed formulas in question. This formula is all about forces released on the tiny solar system to grind the solar system into one conclusion of mayhem and destruction. The inexplicable forces uses the mass of the sun to pull the mass of every and each planet to go towards the sun apparently or supposedly to the tune of the force that pulls across the square of the radius existing between the planet (each to his own) and the sun forming the centre point. What this implies is that Jupiter is rushing towards the sun a rate 317 times faster than the earth and this means by now and after the 4.5 billion years that they say is what the age of the solar system is supposed to be Jupiter either started off being at Oord's cloud and by now is where it is at present but going at a thunderous pace still lapping all other planets on its way towards the sun or Newton's ideas are completely misplaced and totally fraudulent. It has never been determined at what speed is each planet rushing towards the sun notwithstanding that this information will determine the end of the solar system and the end of life in the Universe! Why not?

Then it is expected of me to applaud this corruption because I have to believe in Newton more than I could ever believe in any God and I have to cheer and not jeer when I here the name Newton! Then we get to the real absurd part of the Newtonian religiosity. We get to where from it all started...the comets.

The picture above shows a normal cycle of any of the planets orbiting the sun.

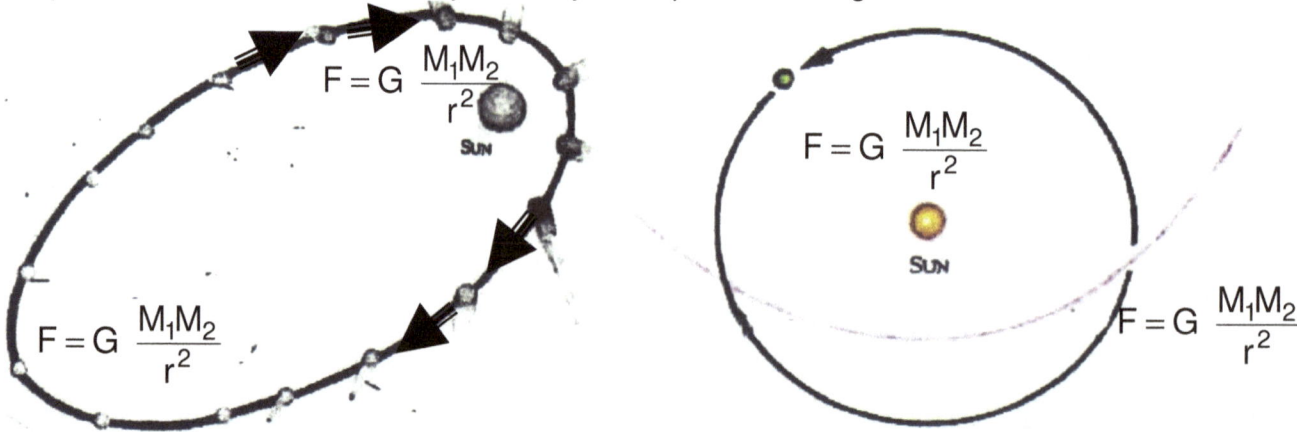

In the case of the comet we have a huge sun with lots of "mass" and a small comet with little "mass". The comet is hiding from the sun where the sun can't see but because the huge sun has lots of "mass" and the comet has little "mass" the sun gets the comet any way. When the sun gets hold of the comet with little "mass" it takes it by the scruff of its neck and drag it all the way to the sun. Because the comet has little "mass" and the sun has lots of "mass" the comet is coming down to the sun because the sun is too big to go to the comet. Then we find the genius of Newton proving to come true. The comet comes to the centre of the sun travelling at a neck break speed to get there. The comet aims at the centre of the sun just like Newton's formula predicts. The formula predicts there is a surmountable force "pulling" the "mass" of both objects by demolishing the radius in the square so that a collision becomes unavoidable due to the force this creates.

$$F = G \frac{M_1 M_2}{r^2}$$

Then as we applaud Newton's brilliance things begin to go wrong. The predictable comet becomes unpredictable. The comet starts to go hay why as it is misbehaving as if it is rebelling against Newton's wishes. The comet should according to the formula head straight to the centre of the sun. It is coming to the centre of the sun as Newton said it would but as it comes close to the sun it misses the sun by a country mile.

The comet redirects the aim and settles on a position south of the sun. This is the first irregular behaviour we find the comet has. Then instead of circling around the sun because of the strength that the small r^2 should provide to the "mass" coming from both sides the comet goes AWALL. …And still the small comet finds the courage to break free from the huge "mass" of the sun. Okay this is where my problem starts with the deductions that Newton concluded. The comet heads back into the dark beyond where the sun can't shine. It never slowed down…it never even acknowledged the grip that the sun's "mass" had so far on the comet. The comet is going back to where it came from without slowing down in the least. But…how and why did the comet move by "mass" because what did mass do on the way down that mass didn't do on the way back. If it pulled then it couldn't push

This is my problem and why **I DO NOT UNDERSTAND NEWTON**. If it was the mass that brought the comet to the sun what is it that pushes the comet back into the dark abyss. What motivates the comet to go back when the mass of the sun and the mass of the comet brought the comet all the way from that dark beyond to the sun. Not one Newtonian master could inform me about this anomaly during the past almost fifty years. All they said was that I was incoherent in thought and that I would never be able to **UNDERSTAND NEWTON**. What is there to understand when I apparently can't understand what pushes the comet away when it is supposedly the mass of both that pulls the comet by the square of the radius to the sun. Then I with me asking these questions I am accused of being a troublemaker that has to be avoided because I am incoherent and my language expertise in English is not up to standard. Because my English ability fails me I have the disposition of being able to follow Newton and find correct conclusions. I have it on paper from UNISA (a university in South Africa) that my English is too poor and that is why I am without any ability to understand the explaining I would need to follow Newton.

Well almost sixty books later you will be able to judge my English and my rejection of Newtonian science.

You the reader, are you able to understand Newton and explain this total annihilation of human logic? This brings a story to mind…there once was this King that bought magic clothes. But there was a catch to the magic. Only the wise and the intellectual superiors in the Kingdom would be able to see the clothes. Those stupid and uniformed, those backwards and common would not be able to see the King's clothes. Therefore the King grabbed at the chance to have the clothes tailored so that every one could see how informed and wise the King was and those in his Kingdom could appreciate just how clever King they had. With the clothes his subjects could truly appreciate their brilliant King's wisdom! The King immediately ordered the magic clothes. He had to look his best when parading in front of his subjects in order to show what a brilliant King they had in him. This repeated in the order of the wise and the brilliant where only the Educated – Masters in science holding wisdom no other humans had and they saw Newton's magic forces flying all over pulling whatever closer.

While on the other hand there were a group of intellectuals that called their faculty physics. They had a magic force that was generated by the "mass" of material. They had this alchemist named Isaac Newton that was a wizard in magical forces and he showed every one the forces at work. However it was only the wise and the clever that could follow Newton's magic. Those stupid and uniformed, those backwards and common would not be able to see Newton's forces at work. The brilliant minded saw the forces by "mass" applying contraction even in the event when the planets are not arranged by mass to perform in the "mass" by which they are awarded. Funny thing is that the big planets are located in the centre of the planetary line up and the small planets are close to the sun and far away form the sun. We would think that with the massive "mass" the big planets have they would travel to the sun at many times the speed of the smaller planets and Mercury as well as Pluto would in relevance seem to go backwards when we compare those two small planets' progress. Yet, with Newton's brilliance notwithstanding a certain order in planet locations are pre-determined by nature, which does not match Newtonian wisdom.

Now we get back to our wise and brilliant King. In order to show his subjects the King ordered a parade through the streets of his Kingdom so that the King could show off his magic clothing. The subjects were to stand on either side of the street and admire his magic clothes. As the King passed the many lines of people they all saw the brilliant clothes and sighed in admiration because there was not one that were unable to see the King's magic clothes. That was until a small girl shouted on top of her lungs: "***The King is naked...the King has no clothes***". Then everybody including the King realised the stupid insanity of the entire event and everybody felt ashamed.

Let's get back to our brilliant Super-Educated physicists that saw how Newton works! Again please try to see the resemblance between the two stories and try to see the connection I have experienced. Newton saw "mass" pulling "mass" and only the few that were very wise amongst the many stupid had the ability to "**understand Newton**". The rest of us were so stupid we could **"not understand Newton"** and therefore were sent home as academic failures because we were "**unable to understand Newton".** How does the comet return to the dark beyond when the mass of the sun pulls the comet towards the sun? By my not understanding this **"I am unable to understand Newton"** or so everybody in physics keep telling me for nearly fifty years. How does "mass" pull "mass" when everything falls equal when under the same conditions but with a variation in "mass"? When I ask this question I am told life is unfair because not everyone is gifted with the ability **"to understand Newton"** and I must except that I lack the mental powers to be able **"to understand Newton".** Thus I am too stupid **"to understand Newton"!**

Now this is one example of Newtonian wisdom that I just don't understand! In light of this conundrum that Newton left us with I wrote a book expressing what must be the most serious issue we on earth must struggle with. **This is the most damning issue that we earthlings have to deal with.**

There are books that can be downloaded for free from the Internet by following Peet Schutte @ Google. In some of these books I expresses the fear that all humans must fight. There should be no other study done on climate change and global warming, not before this issue is very clearly understood and dealt with. We are in immanent danger and we are unable to escape out feat. We are doomed by our being place where we are alive! Every student in physics should make it his or her first priority to solve this riddle before tackling any other field of science. I cannot for any sane reason see how science can priories any study above this one.

We have the mass of the earth pulling on the mass of the moon. Then to complicate matter even further we have the mass of the moon pulling right back on the mass of the earth. All the unknowns are known in $F = G \dfrac{M_1 M_2}{r^2}$. We have the mass of the earth. Then also we have the mass of the moon. We know the distance between the earth and the moon. Newton says vividly this lot is pulling one another and by FORCE no less, closer to each other thus diminishing the distance there is between the earth and the moon... **So when is the earth and its moon going to collide?**
I have searched all over for years on end and I have not found ONE study on this matter. What are they hiding? On who's instructions are they not completing any studies on this issue. When is it doomsday for the earth as its moon destroys the very fabric we call home, our beloved planet.

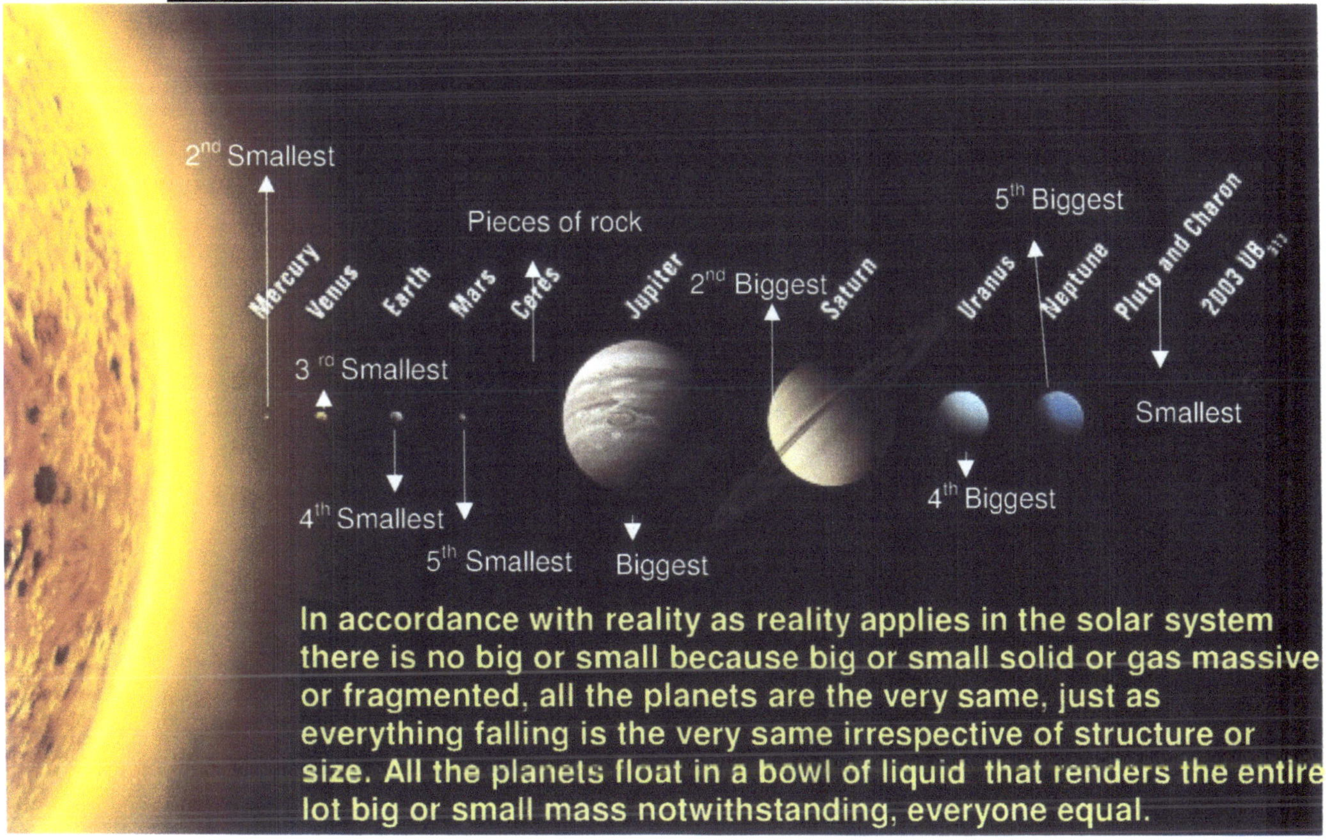

2nd Smallest

Pieces of rock

5th Biggest

Mercury Venus Earth Mars Ceres Jupiter 2nd Biggest Saturn Uranus Neptune Pluto and Charon 2003 UB₃₁₃

3rd Smallest

4th Smallest

5th Smallest Biggest

4th Biggest

Smallest

In accordance with reality as reality applies in the solar system there is no big or small because big or small solid or gas massive or fragmented, all the planets are the very same, just as everything falling is the very same irrespective of structure or size. All the planets float in a bowl of liquid that renders the entire lot big or small mass notwithstanding, everyone equal.

When I ask how does "mass" pull "mass" when the "mass" of planets are completely at random and no selection is obvious according to "mass" then I am told I am of the lesser blessed persons on Earth lacking the ability **"to understand Newton"**. Look at the positional arrangements we have in the planets' formational line-up and the assortment that proves many things but a selection according to "mass" is very much not one of it. However the brilliant physicists ruling the minds of all students in science fortunately and to the luck of all mankind do **"understand Newton"**. How they are able **"to understand Newton"** is unfortunately not in my field of intellect because apparently I am too sparsely gifted. I have tried many times and asked numerous physicists to explain to me these questions but I have not come across one that was blessed to the extent of finding the ability to explain these questions to me. It was forever my fault that I was too stupid **"to understand Newton"** and that rendered them (the gifted ones) unable to show me how **"to understand Newton"**. I was to blame for not seeing the magic of Newton's forces and I was incapable or not wise enough to enjoy the magic of Newtonian wisdom. Are you able to see the magic of Newton or are you just as I am: too stupid to see Newtonian magic.

There was this King and later on there were these physicist…can you see the resemblance in the story? In each case there are only the wise that can see the magic and the rest are too stupid to see.

There are four phenomena that Newtonian science hardly ever mentions but dubbed "a freak of nature". They can't explain these principles but they CAN (not) explain Newton or so they say they can!

There is no evidence of Newton applying in any form. The only presence forming nature is:
The **Titius Bode law**
The **Roche limit**
The **Lagrangian points**
The **Coanda effect**
Then closer to home we also encounter these gravitational laws influencing our every day life.
The entire concept of flying or moreover movement of the earth and things on the earth depends on
The Titius Bode law (forming a ratio that science think of as the Doppler Effect)
The Roche limit (forming the concept we know as the sound barrier)
The Lagrangian points (forming movement limitations in the atmosphere)
The Coanda effect (forming the layers that provide the different atmospheric levels or density levels)
These four laws are crucial in our ability to understand the forming of gravity.
The four are as follows:

The **Titius Bode law** has been around for centuries and with all the mathematical splendour available there for all to use, all the brilliant mathematicians could never come close to show any ability of understanding any of this very important phenomena. They could mathematically equate the formula the sequence applying as the formula, but then after that their superior human intellect dries up as they hide behind worthless equations.

The **Roche limit** has been around for centuries and with all the mathematical splendour available to apply in order to fathom concepts behind this phenomenon, still with all the computing ability of a machine all those physicists with all the mathematical superiority could not touch any understanding about the concept forming the background. Yet when using the truth about gravity in physics the answer is simple; it is that gravity is Π.

The **Lagrangian points** have been known to science for centuries and with all the mathematical splendour available not one calculation could ever explain why this event is taking place.

The **Coanda effect** has powered turbine engines and aeroplanes in flight for almost a century and with all the splendour available to design the most terrific aircraft, not mathematically compute one fact to show understanding why How sad it is that those claiming of much superior intellect in no more than having computing power. The understanding is

mathematical one engineer could this takes place. physics remain just not complex.

These four phenomena or as I renamed them the Four Cosmic Pillars forms gravity. These four phenomena are applying now at this moment while Newton is not present in the solar system. Again I repeat there is no evidence of Newton being in nature! However these four principles are what apply at this moment in nature and has been ruling gravity since the first moment the Universe became a reality. If you deny the existence of these four phenomena as being a freak of nature then you are denying reality as Newtonian science have been doing the past three hundred years.

The explanation or description of the Titius Bode law is as flows and this is taken at random from the Internet by going to Google:

On our solar system scale, an apparent pattern stood steadfast until 1846 was the Titius-Bode rule. This rule noted that the distance of the planets from the sun seemed to follow a pattern described by the equation a = 0.4 + 0.3 × 2ⁿ where n was the planet number in order of distance from the Sun. Then the answer divides by 10 to install the sequence. This pattern held very well for the first 7 planets, so long as one included the asteroid Ceres, or the asteroid belt itself, as planet #5. Yet the discovery of Neptune and Pluto discredited this pattern as a mere coincidence, mathematical happenstance and numerology, as the Titius-Bode rule severely under predicted their distances.

Some still wonder if there wasn't something more to the rule and orbital resonance's didn't have some sort of subtle effect that was being overlooked and made the rule more of a law, at least for innermost planets. With the rapid discovery of planets around other stars, astronomers are once again looking to see if there might just be some sort of truth to this pattern.

Please take note: I have discarded accuracy for simplicity in this following example.
The sequence of the Titius Bode law is to give a planet in a line the value of 3 that doubles to 6 at the next planet and doubles to 12, then 24, 48, 96 and so on. I explain why the doubling happens as part of my explaining the Titius Bode law and a law it is without doubt.
It starts off by giving Mercury the first value of 3 that then places the added value of 4 on Venus.

Mercury. Venus

3 + 4 = 7 divide by 10 is .7

Now Mercury falls away and Venus receives the value of 3 that doubles to 6 and earth is 4

Mercury Venus • Earth

6 + 4 = 10 / 10 = 1

Mercury as a recognised point in the Titius Bode law falls away and Venus then inherit the value of 3 doubling to form 6. The earth then receives the added 4, which makes the compliment 10. The 10 as a value then gets divided by the factor of 10 in the ratio and the earth hold a value of 1

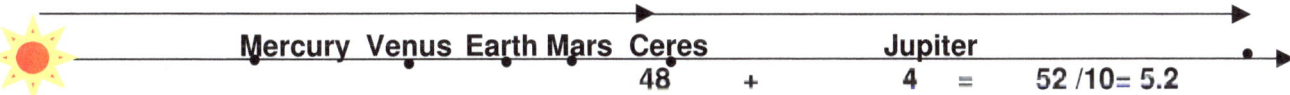

Mercury Venus Earth   Mars

12   +   4   =   16 /10= 1.6

At this point earth doubles the value of 6 to go to 12 and Mars then receives the value of 4. The compliment forms 16 and when divided by the ratio of 10 Mars holds a ratio position of 1.6

Mercury Venus Earth Mars   Ceres

24   +   4   =   28 /10= 2.8

Mars doubles the value of 12 going up to 24 and Ceres / Asteroids forms the value of 4. This total to 28and completes the 28 in he ratio. Then by dividing the value of 28 by 10 the answer is 2.8

Mercury Venus Earth Mars Ceres   Jupiter

48   +   4   =   52 /10= 5.2

This does not even begin to explain what I prove inside. This law is so much more than meets the eye and even life depends on the application of this law. Newtonians, please show me where Newton is in the solar system or show me how I am wrong by discarding Newton to favour nature. This is how the line up grows. Notwithstanding "mass" and irrespective of size the line up we find in nature is according to this ratio. This is what is forming the solar system and Newton is nowhere in sight. In fact Newton plays no part in anything in nature and Newton is just one big gimmick Newtonians put forwards because they have no idea what and why nature holds a total different cosmic compliment than that which they understand. Go to the Internet and research the sequence and you will find it is exactly as this line –up shows. Every time I tried to present the true natural laws applying I first have to completely disqualify anything Newton said because Newton holds no true meaning in nature. Every time I disagree with Newton I get automatically disqualified and my work gets rejected without any Physicist or Academic trying to read anything about my work. Whatever I ever say, it was never good enough to convince Physicists to investigate my work or my study material. They always get away with ignoring my work!

Now I decided to put it in as a prescript and show every person that investigate the book why I am going against Newton. I am NOT going against Newton but I am just following nature while nature is going totally against Newton…and for that Physicists have crucified, humiliated and belittled me for more than twenty years. In fact that was my motivation as to begin to write my 1 book in 1998 and since then it remained an unbelievable struggle to get past the guards guarding physics. That is why I maintain there is a Conspiracy in Science in Progress, which no one can deny.

By **introducing these four laws** the Universe started and I show how that happened as the Universe grew from one point to all the points there now are. I show how the Universe started when it started as one point. I was able to trace the very first point from where the entire Universe began to develop and from there I could determine the centre of the Universe!

To find out where the location of the centre of the Universe is located, then just read further.

This is a warning to those Bible punchers that create their own God by reciting many Bible texts and think by learning the Bible as much as their brains can store they then can through that such reciting get wise: this is not the way this book works. I show how nature proves God and through that you need insight into physics. God created the Universe not by the Bible but by physics. However this is not the book that brings final proof about the Biblical Creation because for that proof I have a selection of other books that refers to the Creation by name. There are those who believe and then there are those who have to understand to believe. This book aims to satisfy those who have to understand in order to believe.

However, understanding is a mental ability and not an emotion as believing or biblical reciting is. This is science and not theology and to understand this requires intellect and not just rehearsing a few verses that you learned off by heart as it is taken from the Bible. To understand this concept I introduce will require a fair bit of study on your part. It is not going to be easy but it will be rewarding.

This is to inform you about my latest book aiming too bring justice to nature by showing how nature and not Newtonian principles form the cosmos. There are four cosmic principles that nature applies and Newtonian science ignores. Nature form the cosmos by applying the following four principles found in nature: These are **1)** The Titius Bode la**w, 2)** The Coanda effect, **3)** The Roche limit, **4)** The Lagrangian points and these four still form everything. These laws are what nature put in place instead of mass and moreover being in place of what Newtonian fantasy makes every person believe is science. To be able to achieve what the Bible says happened I had to achieve what no one in science this far could or would or even tried to achieve in science. I had to break a code of silence.

Before I prove how the cosmos started I first have to prove how nature forms the Universe. This is how nature forms the Universe and these laws form the Universe. From deciphering how these laws apply I could decipher how the Universe started and guess what; I then managed to use mathematics to support what the Bible says how this Universe came about.

This is the Titius Bode law

The Titius Bode law proves that mass has no place in science. See in the picture how random mass is and with such randomness, how can mass place planets in the positions they hold? By my effort to solve the mystery of the Titius Bode Law, I prove that gravity forms not by mass but gravity forms by π forming in movement π². Solving the Titius Bode Law and proving from that how gravity works opens up a new view on the cosmos.

From mathematically explaining how the planet formation forms I could gather the way the cosmos formed right in the very beginning when the Bible said it started the way the Bible said it started. Mathematically proven it is precisely as the Bible says the cosmos started.

This is The Roche limit

The Roche limit has been around for centuries and with all the mathematical splendour available to apply in order to fathom concepts behind this phenomenon, still with all the computing ability of a machine all those physicists with all the mathematical superiority could not touch any understanding about the concept forming the background. Yet when using the truth about gravity in physics the answer is simple; it is that gravity is Π.

This is the Lagrangian points

The Lagrangian points have been known to science for centuries and with all the mathematical splendour available not one calculation could ever explain why this event is taking place. The satellites form precise locations positioned around the major planet and never comes closer while remaining in their positions.

I introduce these four laws as I introduce you to science in the manner as nature forms the Universe. These four laws shows how and even why the Universe started because these four laws are the cosmos and the way the cosmos started. There is no Newtonian trickery. He evidence you read you will make able to understand how things began before anything began because it is nature that brings the proof correctly as everyone can witness from facts that are in nature.

The four cosmic laws hold the basis of what forms the Universe where every one of these laws form part of what forms Π and together these laws form as a group the value of Π^2 and Π^2 is gravity. That I prove mathematically.

The Universe forms the value of 1. No more, no less because that is the value of singularity.

In the beginning there was possibilities before material formed groups that formed material in groups that formed groups of marterial, which we then gave individual names. But before that there were spots and dots as there now are spots and dots forming a Universe. To understand how this lot formed we have to go back and find out how the spots and dots formed before the Universe formed our Universe we give so many names to. Essentially only singularity forms a Universe so small it is outside the Universe we recognise.

This is the Coanda effect

The Coanda effect has powered turbine engines and aeroplanes in flight for almost a century and with all the mathematical splendour available to design the most terrific aircraft, not one engineer could mathematically compute one fact to show understanding why this takes place. How sad it is that those claiming of much superior intellect in physics remain just no more than having computing power. The understanding is not complex. I have to warn the readers that the topics are showing a very new approach with no quick answers. Understanding is in the proof and that does not come by reading just a few lines and then forming conclusions. The information is new but not hard to grasp. I did not put these phenomena in place and these phenomena nullifies Newton's correctness and the proof I bring goes beyond any doubt. I prove the Titius Bode law. Go to the internet and see how science doubt the Titius Bode Law and the correctness thereof while to solve the problem you add 3 plus 4 to get 7. That is if you want to find a solution. I have published the Titius Bode Law in four already published books but in this one I go deeper than the four already published. In each of the books I present I disclose how the Titius Bode Law forms gravity

Everyone is in agreement with Albert Einstein that the Universe started with one spot, a point we call singularity. Now that you saw the laws I wish to give a glimpse about how the Universe started. As I said, the Universe started with 1 and this fact even Newtonian science accepts. Singularity started as the first spot and not with a massive already formed Universe that only afterwards grew in size. Since the Big Bang event the Universe only grew but everything that is in the Universe already was in the Universe. What was in place at the Big Bang only then grouped and became material but not more of what already was. Singularity is eternal without start and without end because when it starts it can't end.

Think of it this way. The spot never changes and was eternally the same. Then one instant the spot overheats and expands into the dot as the movement comes by the spot enlarging to form the dot and by growing into the dot the movement enlarges and the movement as well as the increase in size reduces the heat the dot accumulates. The growth then by increasing in size reduce the heat that cools off the dot too return to being the spot and the cooling removes the structural size of the dot as it returns to form the next spot. Remember what's in the Universe can't leave the Universe but has to remain a component within the Universe as long as the Universe exists. However before heat brought about the Universe we know time was one continuing everlasting spot, which is a line that never went further than one spot.

The spot formed the future. The spot formed the present. The spot formed the past. Since the spot in the future was an exact image of the spot in the present and that was identical to the spot in the past the new spot had no identifiable difference between the future and the one in the past. It was a repeat of what was being identical to what was coming and therefore it stayed the same. The spot was so small it was invisible and unnoticeable and yet it is so big that at present time it holds the entirety of what is within. It is as big as nothing could ever be and it was so small it was nothing that could ever be. This was the only time nothing had a validation because as soon as the Universe came nothing disappeared and became something. One should be very clear about understanding this aspect. The spot then represented nothing because it was consistent of nothing since only nothing existed at the time. It was infinity that which can never start united with eternity, which is that which can never end. To start that which can never end first had to end and start that which could never start and to end it first had to wait for that which can never start to find a way to start to lead to the process of ending. Since neither process could apply due to unfavourable circumstances making it impossible to follow on the other the process was eternally infinite.

The Universe started with the number 1. It is called singularity. Einstein had the idea that this was only applying at the beginning when the Universe started. Thereafter the value of singularity disappeared from the Universe leaving no trace thereof except in some mathematical equation. With this idea there is one big problem and that is when anything becomes part of the Universe it can't go anywhere but stay in the boundaries of the Universe. Thus we have to search for singularity within the Universe and not look elsewhere. With all the space we find throughout the Universe where will we find singularity?

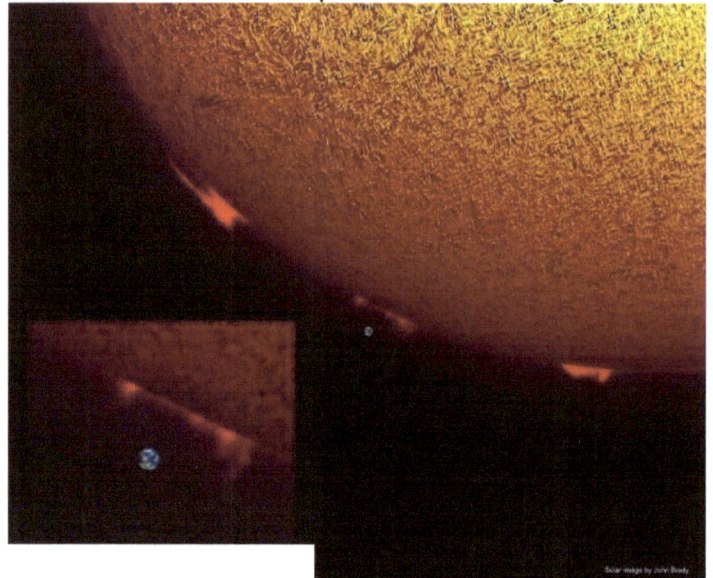

When looking at the sun or the moon or the earth we see a sphere and a sphere is a combination of many circles. When taking the circle we see at all material in the Universe forms as circles that becomes spheres.

Therefore it would be advisable to look at the circle to find singularity. A circle is Πr^2 where Π is form and r^2 is size. What brings about the difference between the sun and then earth and the moon is the factor r^2. In that we find big and small. Looking at the line-up that the solar system forms we can clearly see that big or small plays no role in the cosmos. The planets line up completely random in size and they all form their positions in relation to the Titius Bode law. Therefore with size at random we find we must ignore r^2 because it plays no part in cosmology. Then in the formula of the circle Πr^2 we remove the value of size r^2 leaving only Π in form as a value. This shows that the smallest particle we can ever find is $\Pi r^{2-2=0}$ which is Πr^0. To go one smaller than what Π is we have to remove Π as a factor of value. We then have $\Pi^{1-1=0} \times r^0$ that will be Π^0 and $\Pi^0 = 1$. Having Π^0 we found the value of singularity, which is 1.

To indelicate what I did on a circle I'll draw you a picture because this is very important if you wish to realise where the Universe started.

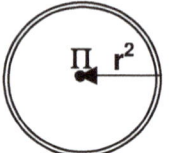

By reducing the radius to become singular meaning it holds no factor in the relation one gets to a point that is a circle. Then getting to the point where we have Π it holds only a value of form but the value of form still has the value of $\Pi = 3.1416$. Then by removing form as $\Pi = 3.1416 / \Pi = 3.1416 = \Pi^0 = 1$. Only then do we reach singularity. In singularity there is no space. As you enter the circle Π^0 you also leave the circle Π^0 on the other side. On the one side all space will move to the right whereas on the other side of the circle all space will move to the left. No person or material can enter this area because there is no room that could hold space. The Universe starts at Π and material within the Universe starts at Πr^2. At Π the Universe we find our bodies in starts. Therefore the point Π^0 cannot be part of the Universe because it is too small to fit into the Universe. This is the point where the Universe started when the Universe started with 1. This point is within every spinning object and because this point holds the value of $\Pi^0\Pi$ the entire Universe notwithstanding space differences is the same size everywhere at all time. This point connects to all points formed by spinning material and because this point Π^0 is the same point everywhere this point connects the entire Universe to the rest of the Universe.

Any person trying to mathematically find a formula that holds a value of anything larger than Π^0 firstly has no idea about what the Universe is and secondly is so far off the mark as far as reality goes they are in the fantasy of Newtonian science where they an create any fantasy because whatever they conclude is just another fantasy. Anything used in cosmology such as $A = \left(\dfrac{E_T}{\rho_0}\right)^{\frac{1}{2}} \left[\dfrac{3(\gamma-1)(\gamma+1)^2}{4\pi(3\gamma-1)}\right]^{\frac{1}{2}} \equiv \dfrac{2}{5}\left(\dfrac{E_T}{\rho_0}\right)^{\frac{1}{2}} \xi_0^{\frac{5}{2}}$ or as $\dfrac{\partial f}{\partial t} + \Pi\dfrac{\partial f}{\partial r} + \dfrac{\theta}{r}\dfrac{\partial f}{\partial \theta} + Z\dfrac{\partial f}{\partial z} + \left(\dfrac{\theta^2}{r} - \dfrac{\partial \phi}{\partial r}\right)\dfrac{\partial f}{\partial \Pi} - \left(\dfrac{\Pi\theta}{r} + \dfrac{1}{r}\dfrac{\partial \phi}{\partial \theta}\right)\dfrac{\partial f}{\partial \theta} - \dfrac{\partial \phi}{\partial z}\dfrac{\partial f}{\partial z} = 0$ has no value at all because the number 0 has no validity as a value and the formula says nothing. Calculating the cosmos is $k^0 = \dfrac{a^3}{kT^2}$

To those in astrophysics this picture represents the cosmos. This is what they believe is out there in space. They do not see reality in nature but to them reality is vested in numbers and formulas of their creation. Nothing is wrong with that except their reality is void of what is reality as far as what we find in nature and that part that forms science I prove is totally unacceptable. From the start they had no idea

about the Universe or about physics in the cosmos. The proof of this statement is in how far their Newtonian concepts are off course with nature. On these uninformed misconceptions they fantasized without limits. I introduce the Cosmic Code but I use nature to prove every aspect of what I claim is the Cosmic Code.

Then on not understanding nature or knowing about nature they became smart by creating fictional mathematics. The biggest person to take the blame about that is Isaac Newton. Let any clever cosmologist prove Newton's formals. We know what the mass of the earth is and we know the mass of the moon and we know how far the two are apart. Let them show when is the moon going to collide with the earth since the two has been pulling each other for billions of years. That is what Newton said happens and they stick to Newton like baby shit to a blanket.

It is said that computers can and will take over man's functions but what is man's functions. If it is to dream up a non-existing cosmos made up by people trying to make fantasy -mathematics apply in a non-functional Universe then yes one may compute computers to become as ridiculous as science currently is. However I challenge any computer programmer to get a computer to figure out how the Titius Bode law works as I did. Figuring out how the 4 cosmic laws work. That is human's actions in applying human brainpower and thinking, but applying non –complying mathematics in a non-existing Universe that is void of truthful and meaningful information is not what honest humans do and to figure out how nature works, well that is exactly what the scientists of the day are unable to do. They can produce make believe and create nothings to fill a Universe but I found they are unable to even read anything consisting of nature's reality and find out what applies in nature as reality will insist they do.

They pretend to determine how the Universe functions by suggesting mathematical impossibilities that they can't ever prove! They think they are so clever that they can create a Universe that is totally void from God's Universe and we all must be very pleased with their fantasy play. Any one of the masters that holds a different opinion to mine and defend science as it is, take up my challenge to show when will the earth and the moon collide as Newton predicted with his formulas.

We as the rest of the humans are regarded as too stupid to follow what hey calculate and so we must sit back and be told how the Universe works. We are not considered smart enough to even insist on explanations because they are as clever as God and we are the scum they have to educate because we are mindless as we are thoughtless with our meaningless existence. They live on a dimension we do not even know exist because we are too meaningless in thought. Any one trying to deny what I say then explain why no one in science even tried to read my work that I have been sending to institutions for almost two decades.

For three hundred years they have been applying mathematical boogieman tactics by scaring every one with bullshit such as this:
$$\left(\frac{P}{2\pi}\right)^2 = \left(\frac{a^2\sqrt{1-\varepsilon^2}}{\ell}\right)^2 = \frac{a^4(1-\varepsilon^2)}{\ell^2} = \frac{a^4(1-\varepsilon^2)}{a(1-\varepsilon^2)GM} = \frac{a^3}{GM}$$
Prove Newton by applying the values he suggested. We know all the values in this equation by now. So just go ahead and prove me wrong by proving Newton correct. Prove how much is Jupiter going faster on route to the sun than Mercury is while Mercury is just next to the sun. Make sense of Newton!

To start with giant formulas as Newtonians do only impresses other dim wits that can't think and only shows incompetence and arrogance as is the case with Newtonian science. **Mathematics came about as the Universe formed and as the Universe formed it formed mathematics during the process. I show exactly** how did the first spot arrive and what was in place before the first spot that became

one. Π^0 **I show why the first spot came into place and what made it be.**

1^0 1^1 I show what made 1 grow into two and why did one grow into two. This is the most fundamental reason why the Universe started. It started with the fact that one came about that became

two. 1^0 1^1 1^2 I show vividly what changes took place to allow two that was one to become three and why did this process begin to form a Universe.

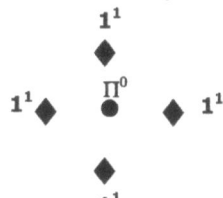

1^1

        Then also why did four become in the place of three without replacing one or two, as a number and I will give a hint; it was because 2 + 2 = 4 then became 2 x 2 = 4. This brought the Universe into a new era such as was never seen before.

Then that which can have no inside parted from that which can have no outside. $\Pi^0 = 1^0$ parted from 1^1. The spot became dot and then relevant as the Universe transformed. Another hint is that this is what the Universe are made of and this is how the Universe formed and transformed into what it became. To claim proof I have to show that this indicated that the motion produces the space and the space finds limits in the motion confirming the space while the space is conforming the motion. In Singularity it is where the

triangle and the half circle and a straight line are equal in 180°

Realizing why these three forms are equal unlocks the information in Genesis 1 verse 1. This was when singularity as 1^0 parted from 1^1 that the motion then came about as 1^2. That in reality left little consolation because with $k^0 = 1^1$ that left the space formed by the motion way outside the realms of the emerging Universe. I decided to replace the symbol Kepler used of $k^0 = 1^1$ to a more appropriate Π^0. However I am

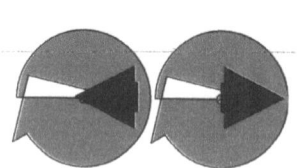

not going to go into detail before I go into detail about what the four cosmic laws are about and why they in movement control the Universe. The concept is confirmed in the fact that by using a pendulum in a time device such as a clock then it proves that time is movement and gravity is movement and that is why gravity moving the pendulum arm is able with that movement to measure time.

Then the Lagrangian law came in place and this allowed four to advance to five. This principle, as is the case with all the principles are still applicable in nature. However how did the development of gravity take one dot to four and then shifted it to five. This is where the

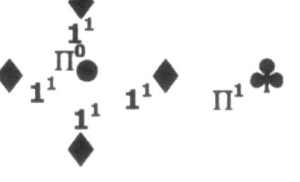

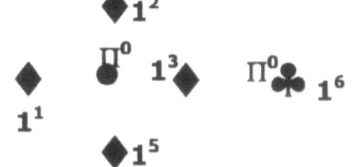

Universe formed $\Pi^0\Pi\Pi^2$ for the very first time.

Then as a result using the reason why two became three then applying the very same reason, it came about to bring five to form six.

However the reason why four became five is the same reason that brought seven into place because the one is the result of the other forming and the two being five and seven are so well interlinked it had to

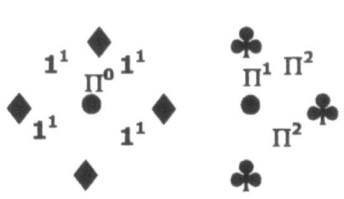

come into place simultaneously. Therefore this then brought about that gravity formed space as $\Pi^0\Pi\Pi^2$. it never formed Π^3 because light brought about that.

By forming seven it then formed ten and that was where that which was eternal and perfect became temporary and imperfect. Singularity distorted into a concept outside the eternal. The truth is at that point all there was, was an inside and an outside still to form and a promise of what might come.

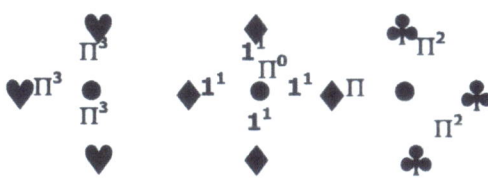

The mighty wise atheists and atheist minded can push their present into forming the past by clinging onto the worthless they represent as worthwhile that will become the past when these books becomes the present. However, that choice will doom them into the past along with all other things and thoughts not worth the burden to take into the future present and onto the future from that past. Their adopting the worthless and not adapting to the truth moves them to the past only worth to be forgotten as the worthless part of the past. The choice is theirs to make.

The mathematics I include is to show what mindless clots those Superior Humans are that portray their position as superior intellectuals in mathematical ability and it is there to disprove the Members of the Physics establishment that advocates the necessity of bringing mathematical proof to prove? I show mathematics not to scare readers away but to silence the Brainy Bunch critics by showing them as being the stupid ones and show the foolishness of their arguments. By using mathematics the Brainy Bunch have been cheating the public and have been brainwashing students for centuries. That cheating is how they do it and I have to show and uncover the dishonesty in mathematics.

The Universe consists of gravity that forms by the working of the four phenomena never mentioned. That does not say much for the bountiful prestige that mathematician's claim as their lawful bragging rights in areas where true human intellect is called on. Is it not high time to begin to admit you are playing the game of fools with you arrogance about your achievements using mathematics when designing space whirls and travelling to galactica while not even understanding what movement asks for? You do not even understand the neutron and the neutron is compressing density increasing, which is what gravity is, which is what time is, which is what all movement is…that is why the neutron has no mass because mass is the principle coming about where independent movement ends.

You're mathematics could not get you any closer than playing games in a fairy tale Universe using misguided presumptions about mass forming gravity and living the Universal farce which Newton created because that fairy land is what all the Kings clever heroes and all the King's splendid wise could never prove in hundreds of years. If you feel superior as a scientist practising physics on the highest level having a gloating hail of superior mental capability covering you like an aura, then I have very saddening news for you. Your Universe that you created by creating "mass" evaporates before your eyes.

If you have the ability to compute and calculate at the highest level, then look at your computer and see one that machine has abilities as a machine which is equal to you, but it's a manmade machine. Stop playing games by creating fairy worlds making up fairy tales about fairies and little people, mass that can create forces, four of them no less, and come and join the rest of us living in reality that does not need to compute forces and see non-existing planets far, far away In a fairyland Universe in order to be able to not understand what it is that you compute, but to use human intelligence and in that way to understand what only human intellect could ever understand. Then what in the present is not worth carrying into the future as the past being worthwhile? Notwithstanding your mathematical brilliance you completely lack any understanding of mathematics or of physics and we see that in your thoughts.

The spot was one perfect spot that overheated and parted into a dot.

•• Then by overheating the spot split into two being a spot and a   dot.

••• The spot shifted to the past leaving the dot in the present while the dot cooled off and formed another spot as one in the past and one in the present and one coming from the future.

The spot overheated as it expanded into becoming a dot going to the value of Π.

While the spot expanded into becoming a dot it   had nowhere to go because it was still inventing space, which was a concept that did not exist yet.

This   was   when 1^0 became 1^1

From the four cosmic laws we can see how the Universe started. I will allow a glimpse into the process, which I wrote several books about and still I know I have not scratched the surface. It started with one spot that then became a dot through overheating on the one side and cooling on the other side. The dot became two, the next spot and the previous dot, because space interrupted time and in between two instances of time landed one speck of space forming a dot on the one side and a spot on the other side. Remember once anything, even if it is a process forms part of the Universe it has no place to go but to remain part of the Universe.

 From this the Universe started exactly as we find the Bible says it started. It started with Π^0 or then singularity or if you wish, then from a void where there is no space and there is only time in the instant. The entirety we think of non- material space is what is filled with this. It is a spot that is not and while it is not it still fills and maintains one entire Universe. Wherever you look you will find this spot that holds space and forms space without being able to claim space. Material fills as a solid because movement compresses this dot so tightly it forms a solid construction of compressed material while the material forms of a substance that claims no space. Non-material is this same heat but is much less compressed and so because of a reduced density because of the flow of time from space to material we can see the density increasing on the one side and decreasing on the other side. The proof of this we see in Kepler's tables where the Titius Bode law process that material is regularly spaced in accordance with the growth formula that the Titius Bode law is indicative of.

Also the Roche limit at $\Pi^2/4$ shows this limit between material forming and the Coanda effect shows a clear growth of density developing around spinning materials as well as the satellite positions of materials around structures we think of as planets. There is a definite distance maintained between matter and therefore it must be non-material the compresses around the sun and while spinning the sun compresses the non-material because this maintains the heat balance within the sun.

When this heat balance goes array the structure overheats and we then find what is thought of as a Super Nova or an exploding star. Everything in the Universe is reliant on density caused by movement, which causes movement, allowing density to development specific time or specific gravitational ratios.

This left one dot and one spot that developed into the next dot. Where are only the first dot, the mother dot, and the original dot from where everything came. We are within that first spot that became the mother dot and that dot we think of the entirety as the Universe that has no end or beginning.

Then time formed a sequence of the past, the present and the future leaving three dots lined up in a line we think of as time.

.1 = past   2 = present   3 = future

Afterwards space becomes a factor because we see the second dimension develop, as time becomes 2 + 2 = 4 and 2 x 2 = 4. This takes the Universe into a complete new level. That says so much in the original book it took me 86 pages to explain this process alone, but we are not delving that deep into the argument in this work. Asking the question how did the numbers as figures become 1, 2, 3, 4, 5... a question of reasoning how 3 became four and how four became five. The entire quest to find the route is so simple it is just a matter of discovering how the four cosmic pillars form the Universe seen from the point nature indicates the route we must follow.

When the Universe reached two 2's the time came when half the Universe went and the other half came. The half opposed the other half and that which was on the one side was a reverse of the other side. This is where polarization came into prominence. Polarization became the basis of gravity and gravity became movement where movement became time.

On the one side 2 + 2 formed 4 and on the other side with a dimensional discrepancy forming for the first time 2 x 2 formed 4. This is where a circle became a line and a circle became a square as it applied on both sides of the Universe. This is where the Universe became a form that later became a cube as 2 x 2 x 2 formed $2^3 = 8$.

That is how the Universe formed as each number brought about a change in its phase of development.

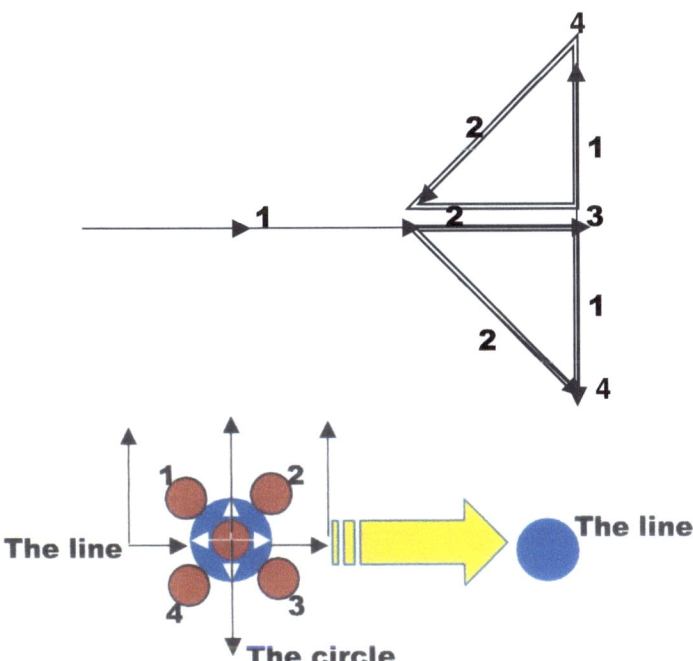

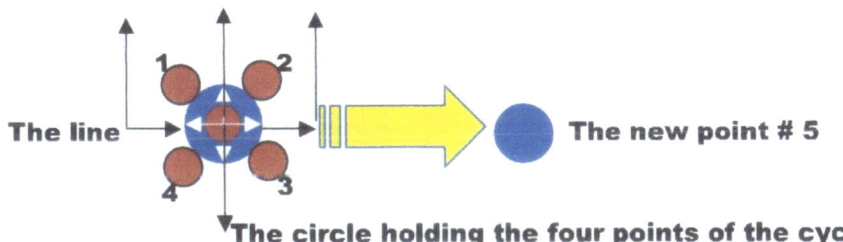

In the one $\{(3 + 4)^2 = (7)^2 + (1)^2 = 50\}$ a line forms a double value and in the other triangle a new point forms the next point in the line.

Both values end up as 10 where the one is $\sqrt{100} = 10$ and the other is $5 + 5 = 10$

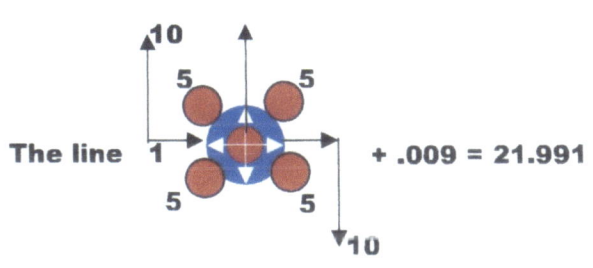

The circle holding the four points of the cycle

However we know that singularity as 1^0 is the smallest point any point can be in our Universe because that point in value falls outside any space within our Universe.

Forming a value as Π of 3.1416 indicates a value of less that 1 which indicates that the line forms a future that comes as the line continues from far outside singularity. In the black hole we can see this clearly where the past eventually absorb and consumes the future into a picture painted in light. The past is the darkness and the future is the line holding singularity.

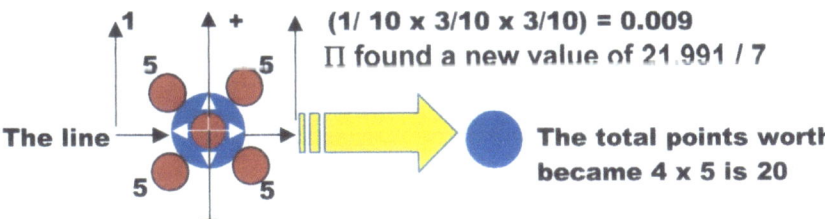

The process in which the circle that holds the five points form a triangle in reference of the circle of four points forming a new cycle.

This is how the Universe began. It began simple as it formed numerical order in numbers that brought about order to spots that became dots. It is so simple that the complicated mindset of the Newtonian physicist rejects this as being too simple for their liking. You will learn not to see the Universe as many but as only one that holds many in relevance of one. However, the four Pillars came in use when the Universe came into tuition. The first three formed time brought in place by the first four that became space where afterwards the first four became five as four brought on five as material.

My personal advice to any potential reader of my writing just please keep the following in mind: My style needs very careful reading: Do not try to speed read or do glance reading but read every word I use carefully and you will do well in understanding because with my writing style it requires much concentration. I don't waste words and I compress factual information.

I now introduce you to the Titius Bode law or how the solar system forms.

There are these four laws or phenomenon or principles in the Universe. Newtonian science can't explain it but I can. In doing so I shatter the myth called Newtonian science and I reveal the hoax science portray for three hundred years or more as the truth. I tried to get published but I was unable because then I break the strangle hold physics has.

Those physicists formulating science are a mafia-gangster club controlling the dishonesty that formulates science… I am one person trying to correct what is a joke but what is also sold as the truth.

Science force humanity to accept the hoax Newton founded and science brainwash everybody to disbelieve nature while clinging on Newtonian hogwash. That is mind-control and that is brainwashing!

I have sent this book with six other books to eighty-five publishers and e-mailed this book to thirty something more. I had no response…not from one. This book opens a new era in understanding how the cosmos works…and not one publisher found it interesting enough to publish or to reply reasons why they found no interest in this as a publishing project. No one is prepared to break the hoax and publish the truth…

In this book I explain in detail the layout working process of Titius Bode law.

This book explains the Titius Bode law but what is the Titius Bode law? As I show you later on the solar system does not form as science and Newton try to show but it forms a reality far apart from what those in science try to promote. The Titius Bode law is as follows: if we give the sun forming a circle in a relevancy of **3** say dots then Mercury holds **4** dots. With that the sun and Mercury places a value of **7** on the space between the sun and Venus. With this ratio the earth will be another **3** dots further and will have **10** dots from the sun to the location the earth has. This is how nature places the planets and this totally clashes with what Newton said happens. I am the first to crack the code as to how and why this happens and believe it or not my work is still rejected in favour of **Newton's ideas that never applied in nature and in reality**. In a sketch this is how nature works!

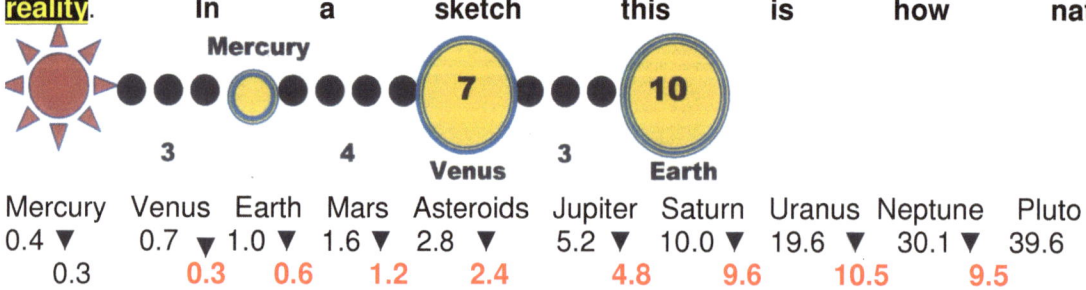

Mercury	Venus	Earth	Mars	Asteroids	Jupiter	Saturn	Uranus	Neptune	Pluto
0.4 ▼	0.7 ▼	1.0 ▼	1.6 ▼	2.8 ▼	5.2 ▼	10.0 ▼	19.6 ▼	30.1 ▼	39.6
0.3	0.3	0.6	1.2	2.4	4.8	9.6	10.5	9.5	

The ratio (not distance) doubles with the placing of every planet except the last two and for that there is a reason. Are you ready to learn the truth about science and see how you were duped? If it is **3** dots from **Mercury to Venus** it is **6** dots from **Venus to the earth** and from **the earth to Mars** it is **12** dots. From **Mars** to the **Asteroids 24** and from **the Asteroids to Jupiter** it is **48** dots with a further **96** dots from **Jupiter to Saturn**. This shows that the distance inexplicably doubles in ratio every time a new planet position is allocated.

For the first time in all of human history there is a method deciphered to show how NATURE no less forms the solar system…and in eighty five DVD's sent plus another (about) thirty six or seven e-mails going via sendspace and not one was interested to publish. Go to the Internet and see it is said this code can't be deciphered but I did find a way to decipher.

That science says is impossible and yet you will read how I did it. Science plainly ignores nature while nature is the reality.

Nature is the only reality but science brushes nature off the table, as if nature is madness. To so many publishers I sent the entire book…I sent it as a unit with two chapters more than the book you read and found no publisher prepared to take on science and correct the hoax Newtonian science is.

Now you can find out how to crack the code by which nature (not Newton's fiction) forms the Universe in the manner that it forms the solar system. It is simple; it is adding 3 plus 4 to get 7! Finding the Titius Bode is 7 also turns what you thought was cosmology into an explanation of the truth while the truth turns cosmology into truthful science. This Titius Bode law, its 7 / 10 or 10/7. There is a ratio. If you give the sun **3** dots then the rest is:

3   x 2 = 6   x 2 = 12   x 2 =   24   x 2 =   48   x 2 =   96

Are you familiar that this is the true state of affairs in the solar system and that mass of planets never played a part in the allocating of any position in the planets' layout? Say, if Venus has 3 dots then   x 2 = the earth will have 6 dots

If the earth has   6 dots x 2 = then Mars will have = 12 dots

If Mars has 12 dots x 2 = then the Asteroid belt will have 24 dots

If the Asteroid belt has 24 dots x 2 = then Jupiter will have 48 dots.

If Jupiter has 48 dots   x 2 = then Saturn will have   96 dots.

How aware were you reading this about this law called the Titius Bode law? How much do you know about the presence of this law and were you aware that nature annihilates everything Newton says with this law. This is how nature uses planetary positioning but those In science totally denies this as a reality because nature destroys everything Newton said with the application of this law. The Titius Bode law is reality notwithstanding those in science trying to belittle nature and deny this law and that is a reality you can look up wherever you wish to look. Please take note of my accomplishment. I am the first one in three hundred years to crack this code. Do you even think that I am clever? I surpassed all genius that came before me and this is why: I am clever because I am the 1st in 300 years that could add 3 (+) 4 and get seven (7).

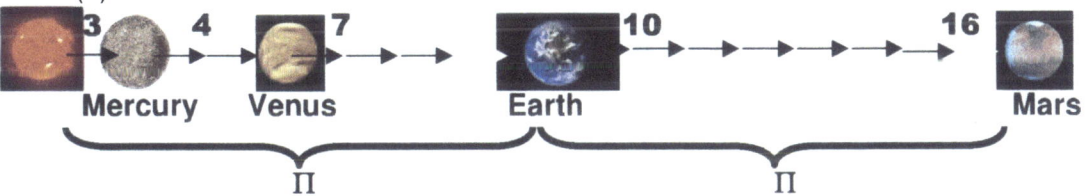

No one in 300 years broke this code because it is so extremely complicated…and you believe that? I am the first person that can add the value of 3 and 4 and get 7!
This is how the sequence work and this sequence have been known to science ever since 1776. When the sun holds an initial value of 3/10 the ratio is as this table predict:

Mercury	Venus	Earth	Mars	Asteroids	Jupiter	Saturn	Uranus	Neptune	Pluto
0.4	0.7	1.0	≈ 1.6	≈ 2.8	5.2	≈ 10.0	≈ 19.6	≈ 30.1	≈ 39.6

Then with Mercury forming the next planet in the order Mercury holds a ratio adding value of 4. Then you divide the total by 10. This is the sequence nobody could unravel.
This is the complicated part. Adding the first **3** and the 2nd value of **4** puts **Venus** in a ratio position of **7**. Then you add **3** to the tally and divide by the **10** the earth has to get **1**, which is the **earth**. The doubling is as simple as I explain inside.
It is so simple as adding 3 plus 4 that then divides by 10. This becomes a mystery no one in 300 years could solve or ever since Newton came up with his idea of mass forming gravity. Do you also smell a rat or is this statement I make going too far? Do you believe this hogwash that nobody could add 3 and 4 to get 7? …And it took the most brilliant minds 300 years not to conclude this. Those in science, the lot are always going on about telling everybody that this order or planet line –up does not make sense and this is nature's (freak) or failure and this is a reject from the madness that nature is. The question is always about the Titius Bode law or planet ratio that could never be mathematically explained or put into context

with gravity. Science always places doubt about nature. It is always putting nature in doubt and then they put Newton's ideas forward as if it is self-explaining and that is untrue because Newton is found nowhere. There is a conspiracy to prevent the truth about Newtonian science to become unmasked. That statement about corrupt science I prove and that is why the order of science rejects all my work. I ask questions because I give answers they do not wish to discover. If you feel offended or hurt or worried when reading this you are brainwashed and duped at the same time. Please, for God sake then put this book down. Do not read further because with your limited vision and understanding of science this will crack all your confidence. You have been mentally conditioned to believe Newton and to accept science as the only truth there ever was and this will come as a shock and destroy you!

This, the Titius Bode law is there. This is undeniable what is in place. Anything other than this is a daydream and seeing other wise then becomes the King's magic clothes, which is precisely what Newtonians think they see, they think they see Newton's magic gravity mass pulling. Newton isn't and nature is this law and anything ells is a hoax.

Mercury is 3 dots

This discrepancy why Mercury and Venus both form 3 I explain in this book.

Venus is 3 dots

This is the relative points in ratio that Venus holds as a distance from the sun.

Earth is 6 dots

This is the relative points in ratio that Earth holds as a distance from the sun.

Mars is 12 dots

This is the relative points in ratio that Mars holds as a distance from the sun.

Asteroids are 24 dots

This is the relative points in ratio that Ceres holds as a distance from the sun.

Jupiter is 48

This is the relative points in ratio that Jupiter holds as a distance from the sun.

Saturn is 96 dots   This is the relative points in ratio that Saturn holds as a distance from the sun. There is no space on the paper provided in this book to indicate the others but Pluto is 40 times further than the earth is. Why do the planets' distance double?

I explain this in detail because what there is in the solar system is this; the Titius Bode law forming in conjunction with the Coanda effect, the Lagrangian points law as well as the Roche lobe / Roche limit that is what forms gravity. This is it!

Newton and his ideology is as absent as the correctness Newtonian science hides and if you do not believe me go and research this by yourself. Giving this truth I contacted (about or more or less) 150 publishers among which there are about fifty or more Universities and not one had any interest in publishing this book. What you are going to read is the result of the first time in human history that any person had the inclination to explain the forming of the solar system and still nobody is interested in this reading about how this venture unfolds or how the forming of the solar system concludes.

This is how space forms and out there amongst all not one shows interest in publishing?

However this is not where the influence of the Titus Bode and other three laws end. Here is just another example showing how this changes our perception of what really applies in science. Furthermore I prove mathematically that electricity and gravity is the very same thing but it applies on different dynamics or on different exponentials.

By using the Coanda effect the earth spins and this spinning causes cosmic liquid to electroplate onto the earth surface thus making the surface expand or grow. All spinning material is getting larger with time.

The Coanda effect works on the principle of applying $\Pi^0\Pi\Pi^2$, which is a derogative of Kepler's formula $a^3 = T^2 k$, which then in essence becomes $k^0 = \dfrac{a^3}{kT^2}$ and I then changed that to $\Pi^0\Pi\Pi^2$.

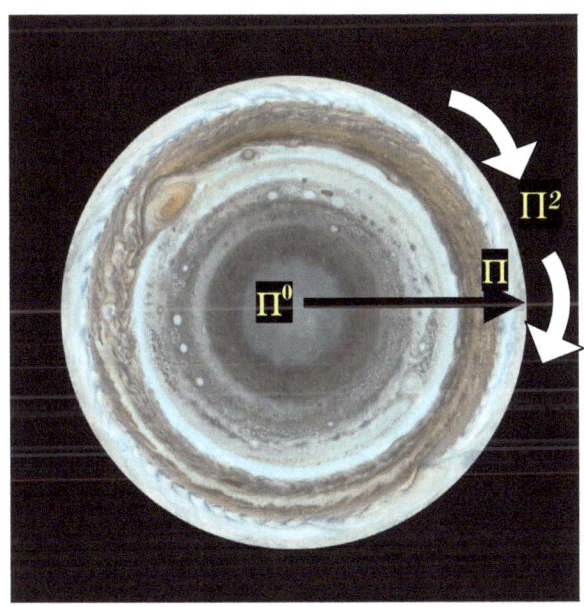

When there is movement between the liquid and the solid such movement establishes a point in the centre of the circle or the sphere that forms singularity Π^0. There has to be movement and according to cosmic law the movement can only be with the liquid. This I explain in much detail. The circle then establishes a limit that has to be round and forms a value of Π. This forms a non-existing line between singularity and the circle edge $\Pi^0\Pi$ however this line only forms because of the movement of Π^2. However the circle has to form Π to extend $\Pi^0\Pi$ because of Π^2

It is the movement Π^2 that establishes the line $\Pi^0\Pi$. The movement will always sort under the liquid while in cosmic law the solid will always stand still ass $\Pi^0\Pi$. That is why we see the sun that rises and everything in outer space turn around the earth's singularity. This made me realise how the Universe started from singularity. The faster the liquid goes in movement Π^2 the longer would the line $\Pi^0\Pi$ be in relation to the movement extending the influence of the circle, thus densifying the liquid to a solid.

This process is not recognised be mainstream as gravity but this is how gravity applies. Outer space is a cosmic gas. By turning the earth densifies (a new word created by me) the cosmic gas to a cosmic liquid. In this process the earth grows by electroplating.

The core grows in ratio 56 to the silicon's 28 and the water in ratio grows at a rate of 17. I explain this ratio in the books.

This makes the core grow much faster than the topsoil does and this forms the earthquakes we experience. The cracks open because the topsoil can't grow by the same margin as the core. In that we find that the earth rises constantly above the sea level as everything on earth seems to have been bigger in the past as it is in the present. Going further back we find everything back then was a giant compared to today but that is only because the earth as the entire Universe does expand in every sense there is and what was becomes bigger as it still grows with the earth. This we witness in fossil remains being so big. The fossil remains and even the fossil discard grew with the earth and now are many times larger than what it was in the past. The dinosaurs back when they were alive were the same size as modern lizards but the Earth was just like the Universe confined to much less space than in the present. This influenced nature and the weather and by the weather changing that had a tremendous in the development of life on earth.

Among us the many some of us just believe because we accept and others have to understand to accept truth and you can choose where you are. I cater for both categories by supplying mathematical equations but if you don't like maths just ignore it.

Please be warned about the following:

Reading this book will intellectually find the reading to be very challenging to any person since what I say was never yet published. Everything you are about to read is new! That I in principle disagree with science's accepted principles on very basic issues is a fact that is undeniably true. What you read about the principles I propose is new to everyone alike. However, I found that the ordinary persons with a scholastic physics background cope with the difficult explaining much better than does Super-Educated-Masters. The Super-Educated-Masters have information stored by culture and if they can't bring the information to mind by recognition of it they fail to understand new science concepts. You are going to read this in a letter that was sent to me.

The purpose with which I wrote this book is to get around the network of Super-Educated-Masters who strangle any information that forms of science in the form I propose and therefore that does not fit their views or match their liking. If what anyone says does not stroke with what the Brainy Bunch says who controls physics and agree with "Mainstream Science" or echo their thinking, they just smother all intellectual publication on the grounds that it is not fitting their profile on science. If you ignore culture and read with using your logic you will find many accepted norms as ridiculous. With most concepts I disagree most strongly and I disagree because those concepts lack proof but I do also supply detailed proof of my views and that is where Mainstream Science blocks the publishing of my views on science that does not compliment their views.

Read this and wake from the culture you believe in; that which science has lulled you into and made you accept science as the absolute undeniable rock fast truth by instating it as a religiosity then stop reading or get your tranquillising anti depressants next to you with a large bowl of water and a big glass. You will find some mathematical equations, if you are not familiar with it ignore it because it shows the silliness of "Mainstream Science" but if you don't read it you will still understand the explaining by reading the language where I explain it. "Mainstream Science" hides behind maths. I need help to fight their fraud and I need you to help me fight them. What you read I prove to even every last detail and even in this book and therefore I dare anyone to prove otherwise or reprimand me.

You are not going to read a book but you are going to travel a journey. Everything written in this assembly is collected from numerous articles and papers I wrote to individuals, journals, science magazines, Universities, academics in administrative-teaching positions as well as many on line physicists which I tried to interest in my view and my findings on science. I did not remove parts or sections of the articles to disguise this as a book that is ready-written but formed it to be a road on which I travelled and the way I found never-ending rejections.

I never tried to make a name in physics but tried to get some money to make a living from and take care of my children. I always knew there were so many smarter brains out there than what I have and who could see what I saw and take the challenge of correcting science from what I brought to the table. I was stonewalled by a bunch of corrupt conspirators trying to lay claim to rubbish Newton presented as truth and in some cases their efforts to justify Newton's lies became pathetically poor.

I realised there are those that has brains and then you have those that understand Newton and that is why astrophysics remained backwards as it got stuck promoting forces flying all-over pulling to form gravity. I say this straight: no amount of words can describe my utter disgust I have for those in science thought of as flawless performing beyond blame because they are criminals hiding their criminality.

Why would knowing the Titius Bode improve our understanding of physics?

There are 4 laws applying in nature and that is the part that science ignores because with these laws Nature is Annihilating Newton

This is the Titius Bode law

The Titius Bode law proves that mass has no place in science. See in the picture how random mass is and with such randomness how can mass place planets in the positions they hold. By my effort to solve the mystery of the Titus Bode law I prove that gravity forms not by mass but gravity forms by Π forming in movement $Π^2$. Solving the Titius Bode law and proving from that how gravity works opens up a new view on the cosmos.

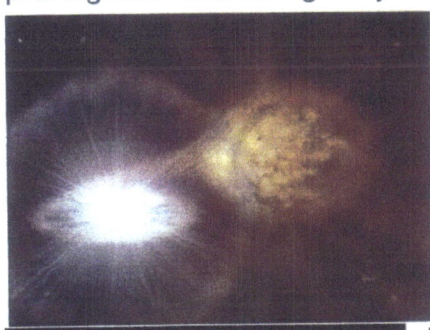

credit: NASA/JPL

This is the Roche limit. The Roche limit proves amongst others how the sound barrier applies and works. It also proves that cosmic structures with an atmosphere can never collide because the Roche limit that produces the atmosphere prevent foreign object from moving faster that $Π^2 / 4$ within the boundary limitation of that atmosphere. The Roche limit brings further proof that using the truth about gravity in physics the answer is simple; it is that gravity is Π.

This is the Lagrangian points

The Lagrangian points have been known to science for centuries and with all the mathematical splendour available not one calculation could ever explain why this event is taking place. The satellites form precise locations positioned around the major planet and never comes closer while remaining the their positions.

This is the Coanda effect

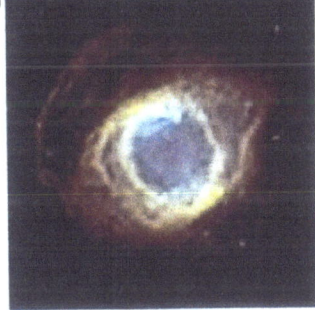

turbine engines and century and with all the design the most terrific mathematically compute one takes place.

How sad it is that those physics remain just no more understanding is not

The Coanda effect has powered aeroplanes in flight for almost a mathematical splendour available to aircraft, not one engineer could fact to show understanding why this

claiming of much superior intellect in than having computing power. The complex.

I have to warn the readers that the topics are showing a very new approach with no quick answers. Understanding is in the proof and that does not come by reading just a few lines and then forming conclusions. The information is new but not hard to grasp. I did not put these phenomena in place and these phenomena nullifies Newton's correctness and that proof I bring goes beyond any doubt. I prove the Titius Bode law. Go to the Internet and see how science doubt the Titius Bode law and the correctness thereof while to solve the problem you add 3 plus 4 to get 7 that is if you want to find a solution. I have published the Titius Bode law in four already published books but in this one I go deeper than the four already published. In each of the books I present I disclose how the Titius Bode law forms gravity

These 4 laws are part of nature. Go look it up before you go on…and why don't science recognize these laws…because these laws brings the entire industry of science to an abrupt end and will stop everything science put in place as science.

Recognizing the importance of these laws will kill an industry worth trillions…

I now introduce you to the Titius Bode law or how the solar system forms.

There are these four laws or phenomenon or principles in the Universe. Newtonian science can't explain it but I can. In doing so I shatter the myth called Newtonian science and I reveal the hoax science portray for three hundred years or more as the truth. I tried to get published but I was unable because then I break the strangle hold physics has.

REASONS WHY YOU SHOULD PURPOSELY READ IN ORDER TO INVESTIGATE AN ACADEMIC INTRODUCING TO THE TITIUS BODE LAW BOOK 1, 2, AND 3

In this book for the first time in **human history** everyone can find out WHY the solar system forms as it does because this book for the **first time ever shows** that the solar system forms by the principle we call **the Titius Bode law.**

The incorrectness in Newtonian science is clear in the way that

The **Titius Bode law** is about the sequence of space every planet holds that becomes bigger the further it is from the sun. Every planet doubles the relevancy in ratio to what the previous that is more to the inside planet and closer to the sun holds and then is half the relative ratio of the planet to its immediate outside. That brings about that the space the planet holds is by ratio doubling and makes the Newtonian idea of mass pulling mass obsolete. Although similar in size Pluto is 100 times further from the sun than Mercury the one is closest to the sun and the other is furthers from the sun. The Titius Bode law forms the solar system by perfectly predictable sequence and that destroys all Newtonian credibility.

mainstream science is hiding the importance of these four laws that

The **Roche limit** has been around for centuries and with proving that gravity forms by movement forming Π whereby I show not only that it does accommodate Π but also why the Roche limit is in place and what influence does the Roche limit have on ordinary gravity applying. The Roche limit is responsible for the "sounds barrier" and it is when movement exceeds the Roche limit in relation to Π that the "sound barrier" becomes a factor. Yet, when using the truth about gravity in physics the answer is simple; it is that gravity is Π. By deciphering these four principles that rule nature I prove gravity is formed as it forms Π.

annihilates Newton. Read what they cover-up when hiding these 4 laws. These

The **Coanda effect** has powered turbine engines and aeroplanes in flight for almost a century and with all the mathematical splendour available to design the most spectacular aircraft, not one engineer could mathematically compute one fact to show why this law applies. How sad it is that those claiming of much superior intellect in physics remain just no more than only having computing power. The understanding is not complex. The entire Universe works on a relevancy that exist between material holding space and material claiming space, in other words the relation there is between solid and liquid space. The Universe comprises of solids with a density beyond the speed of light and liquids below the speed of light. This relationship is where movement established Π and in the way Π forms the density of space revaluing from 21.991 / 7 to holding singularity at Π being 3.1415 / 1. I prove by changing Π from 21.991 / 7 to 3.1415 / 1 it condenses the space surrounding material.

4 laws are placed in use by nature and science ignores these laws outright…but why?

The **Lagrangian Points** I prove by proving how Π forms as gravity. The rings form around planets as the rings form Π. Everything about the cosmos forms by the manner in which Π forms and all movement in the solar system adheres to Π allowing movement Π^2.

In short the Titius Bode law states that **the way we find the solar system forms, this is a fact beyond dispute! There is a sequence in which the planets are spaced and this spacing has no relation whatsoever to mass of any description albeit planets, rocks or distances in formation.**

The Titius Bode law named after Johann Elerty Bode (1747- 1826), who in 1772 published the law, formulated by Johann Titius in 1766 as Astronomy an empirical rule relating the distance of planets from the sun, based on the numerical sequence 0, 3, 12, 24,... By adding 4 to the sequence each time after the numbers doubled and afterwards the dividing the resulting number by 10 results in a sequence of 0.4, 0.7, 1, 1.6, 2.8, which is a reasonable representation of the actual distances in astronomical units for most planets but that is if the minor or inner planets are counted as a single entity at 2.8

In a sketch it explains as follows:

3 6 12   24   48   96   192

The first planet will adopt a value in ratio of 3. The 3 then adds 4 and the result divide by 10 in order to locate a distance in ratio in order of the Titius Bode law.

In the first 3 inner planets the ratio does not fit this explanation exactly but the reason why this does not apply has to do with gravitational singularity, which is a major new concept I bring to science. Then to complete this ratio at every planet's worth in distance 4 is added and 10 divide the number. This is how the solar system forms and there is no other way. The reason why it starts with 3 that I explain by proving the concept... The reason why 4 gets added that I explain by introducing the correct laws... The reason why the distance doubles every time becomes clear with using very simple mathematics.

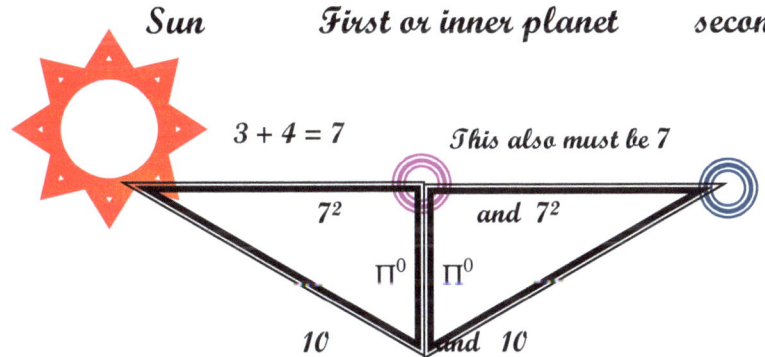

Sun   *First or inner planet*   *second or outer planet*

$3 + 4 = 7$ *This also must be 7*

7^2 *and* 7^2

Π^0 Π^0

10 *and* 10

The law of Pythagoras proves that in the triangle with 90°, then $7^2 + (\Pi^0)^2$ is 50 and 50 + 50 (both being the same) is 100.

According to Pythagoras 100 is $\sqrt{100}$ and in context that is 10.

When the distance from the sun to the inner planet is $3 + 4$ the total has to be 7. In the event of the distance between the second planet and the first planet being equal to the distance between the sun and the first planet it is obvious to presume that the second planet distance must also be 7 and the combined distance must be $7 + 7 = 14$.

The ratio then is $\dfrac{7+7}{1}$ *and* $\dfrac{10}{7}$ *In thee vent where* $k^{-1} = \dfrac{T^2}{a^3}$ *is* $\dfrac{7+7}{1}$ *then* $\dfrac{7+7}{1}$ *is*

$k = \dfrac{a^3}{T^2}$ *must be* $\dfrac{10}{7}$. *With that the case the application of the law has to be* $\dfrac{7+7}{1}$ *divided by*

$\dfrac{10}{7}$. *This argument I prove later on in the book.*

$14 \div 1.42 = 9.8591549295774647887323943661972 = 9.856 = \Pi^2$ *and that is the value of gravity. Gravity might be 9.81 in England, France and Germany but on earth it and in thee cosmos at large gravity is $9.856 = \Pi^2$. This is then what I prove in this book.*

However, this is the obvious where the four pillars form the way gravity conducts control over all movement we endure on earth.

As long as everybody thinks science is more perfect than God Almighty (and I am not trying to be sacrilegious but just truthful because that is the image science portrays) there is no need for the work I present since there then is no need to scrutinise, that which is perfect. As long as everybody goes around with the concept that science is beyond reproach and is as truthful as religion should be no one will take note about the need or indeed the urgency that my work present but science is in dyer need of reforming very serious mistakes. As long as the world regard Physicists as to be as truthful as a Pastor/ Preacher/ or any man of the cloth is suppose to be then no one will think twice to read my work because why waste time scrutinising the already flawless. Therefore in the present and the here and now I am going to show any person that reads this how disgracefully corrupt, inaccurate and disgustingly wrong Newtonian science is and how big a conspiracy they uphold to guard Newtonian flaws. If Science can't show by physics is. Where science declares that there is not a God that created the cosmos then it is put forward that there is no God Almighty because Science is unable to prove a God. With science regarded as everything that is flawless and absolutely correct there is never even given a thought that it is Science that is too feeble and incorrect to prove the existence of a God or Creator and science with the corruption it is formed by, has not the ability to prove God Almighty. No, the existence of God Almighty comes into question and the question is never brought to the door of the accuracy or the ability of Science. This book is about how incorrect the basis of science is and how nature proves Newton's misconceptions. At this point I nailed my coffin and sealed my rejection because going against Newton is science- blasphemy and such a person should be shot. Science hides Newton's flaws by ignoring and disregarding such flaws However science cannot ignore the Titius Bode law, which I prove, and which forms the solar system!
There are many books written by me on the subjects mentioned that is much more informing and contributes much more information as to prove. However, this book serves as a synopsis of other books.

This book does precisely what the title says. This book shows how the principles that nature applies at present and throughout time completely destroy all integrity Newton's principles on science claim to have concerning validity or claim there may be on accuracy or any other form of factuality it may have thought to carry. Newtonian science has no foot to stand on! In the website **naturescosmicconcept** there are information about other work I have that will bring more light to what my work uncovers.

The idea science put forward that nature and Newton fit hand in glove is a myth carried on by mainstream science by never mentioning the differences between nature and Newton. This is ongoing for more than three hundred years but Newtonian, which is accepted science holds as much truth as Newtonian approach to cosmology present in accuracy and that is nothing. Newton's views and claims on physics and what nature applies do not even share a Universe in comparison.

Science and those responsible for formulating science principles know that nature is completely different from Newtonian science and in three hundred years there was not one attempt to correct this anomaly or to justify the difference there is between nature and Newton. I prove this to be a factual deliberate attempt carried out by members in science. Therefore the deliberate attempt is to sweep nature under the table pretending that nature is completely inaccurate, where Newton is portrayed as being the ultimate accurate view there ever was. This misconception is a conspiracy and that I prove so openly it becomes a joke. Nature is shown as a joke and reading a short book I hand out for free will embrace and prove what I say with much clarity.

I broke free from incorrect science by proving nature holds validity in spite of science trying to diminish and nullify nature or the ultimate role nature plays. Because of the implications that this book brings there is no chance that the normal route of Science Publishers will accommodate me and therefore I need another avenue to find publishing either through a publisher or an investor ready to make money. Mainstream publishers on science blocks all my attempts because my work destroys Newtonian science. For the first time in human history I manage to prove nature and the way nature applies gravity. In the past no attempt was ever made to prove nature.

In this book I prove a principle called the **Titius Bode law**. In more complex books I prove the other three as well. The **Titius Bode law** is the way nature forms the solar system and this principle and the way nature applies the principle annihilates everything Newton ever said about cosmology. However, this is not the only new concept it brings about.

The truth about what gravity is shows the Universe in a total new revealing light. When applying physics in accordance with nature and the four principles nature uses to form the solar system, the entire concept becomes new and the entire concept forming cosmology gets a clarity it never had, regarding not only the physics aspect but also the entire picture of how the cosmos forms that is rewritten. I base everything on the findings of Johannes Kepler and the way Johannes Kepler formulated the solar system. Newton in complete ignorance of what science is made such a hash of Kepler in the way Newton raped Kepler's work that read as Newton corrupted Kepler's work it is completely laughable.

This I prove in this and many other books.

This is what this book brings; it brings the new concept that introduces the new clarity concerning cosmology. Everything and every principle applying you are about to read is something you have never heard of before.

The Absolute Relevancy of Singularity The Article free of charge
The Absolute Relevancy of Singularity: The Dissertation, then when convinced about the authenticity buy
Book 0) The Absolute Relevancy of Singularity in terms of Newton
ISBN-13: 978-1530754373 and ISBN-10: 1530754372

Book 1) The Absolute Relevancy of Singularity in terms of Cosmic Physics

ISBN-13: 978-1523722457 and ISBN-10: 1523722452

Book 2) The Absolute Relevancy of Singularity in terms of The Sound Barrier
ISBN-13: 978-1523723027 and ISBN-10: 1523723025

Book 3) The Absolute Relevancy of Singularity in terms of The Four Cosmic Pillars
ISBN-13: 978-1523726462 and ISBN-10: 1523726466

Book 4) The Absolute Relevancy of Singularity in terms of The Cosmic Code
ISBN-13: 978-1523729180 and ISBN-10: 152372918X

Book 5) The Absolute Relevancy of Singularity in terms of Life
ISBN-13: 978-1523729739 and ISBN-10: 1523729732

Book 6) The Absolute Relevancy of Singularity in terms of Investigating Kepler
ISBN-13: 978-1523732074 and ISBN-10: 1523732075

Book 7) The Absolute Relevancy of Singularity in terms of The Theses
ISBN-13: 978-1530582198 and ISBN-10: 1530582199

Book 8) The Absolute Relevancy of Singularity in terms of A Cosmic Creation
ISBN-13: 978-1530648269 and ISBN-10:1530648262

Which all are also available from Lulu.com.
Which is the work I propose that any person should read in the order that I suggest above, should any person be interested in finding out more about what was never yet revealed. Then, before venturing into the printed work, please make sure you have read the recommended e-book titles. What ever you know about cosmology or about physics, you will venture into a Universe you have never encountered before and the logic and the proof that I build on in the printed titles I derive from the proof I give in the e-books where I then do not repeat the proof I claim and which I give in the e-books. By completing the e-books you will be one of a few that truly knows what gravity is! The Absolute Relevancy of Singularity The Dissertation is there written as the second introduction to introduce the four pillars in a very wide sense on which the new theorem rests.

Then The Absolute Relevancy of Singularity consists of a four individual part theses each forming a thesis. There are either six individual books on offer in e-book format or in print could only be purchased as one unit named The Absolute Relevancy of Singularity The Theses

An Academic Introducing to The Titius Bode Law Book 1
Authored by Peet (P.S.J.) Schutte

ISBN-13: 978-1507845851 (CreateSpace-Assigned)
ISBN-10: 1507845855
BISAC: Science / Cosmology

500 pages **$99.99**

An Academic Introducing to The Titius Bode Law Book 2
Authored by Peet (P.S.J.) Schutte

ISBN-13: 978-1507853788 (CreateSpace-Assigned)
ISBN-10: 1507853785
BISAC: Science / Cosmology

An Academic Introducing to The Titius Bode Law Book 3
Authored by Peet (P.S.J.) Schutte

List Price: **$99.99**

8.5" x 11" (21.59 x 27.94 cm)
Full Colour on White paper
500 pages

ISBN-13: 978-1505874884 (CreateSpace-Assigned)
ISBN-10: 1505874882
BISAC: Science / Cosmology

A Cosmic Birth as an Academic Presentation Book 1
Authored by Peet P.S.J. Schutte

ISBN-13: 978-1517066970 (CreateSpace-Assigned)
ISBN-10: 1517066972
BISAC: Science / Cosmology **$99.99**

A Cosmic Birth...as a Special Presentation Book 2
Authored by Peet (P.S.J.) Schutte

ISBN-13: 978-1517525460 (CreateSpace-Assigned)
ISBN-10: 1517525462
BISAC: Science / Cosmology **$99.99**

494 pages

Nature Annihilating Newton: (The Titius Bode law Deciphered)
Authored by P.S.J. Peet Schutte

ISBN-13: 978-1492705666 (CreateSpace-Assigned)
ISBN-10: 1492705667
BISAC: Science / Cosmology

Introducing The Titius Bode law: Nature Works in the Natural Universe
Authored by Peet (P.S.J.) Schutte

ISBN-13: 978-1506194615 (CreateSpace-Assigned)
ISBN-10: 1506194613
BISAC: Science / Cosmology

Cosmic Secrets Decoded Part 1: Proving the Titius Bode law Working
Authored by Peet (P.S.J.) Schutte

ISBN-13: 978-1505866513 (CreateSpace-Assigned)
ISBN-10: 1505866510
BISAC: Science / Cosmology

A Cosmic Birth as an Academic Presentation Book 1
Authored by Peet P.S.J. Schutte

ISBN-13: 978-1517066970 (CreateSpace-Assigned)
ISBN-10: 1517066972
BISAC: Science / Cosmology